John Milnor

Dynamics in One Complex Variable

John Milnor

Dynamics in One Complex Variable

Introductory Lectures

2nd Edition

vieweg

John Milnor
Institute for Mathematical Sciences
State University of New York at Stony Brook
Stony Brook, NY 11794-3651
USA

1st Edition 1999
2nd Edition 2000

Vieweg is a company in the specialist publishing group BertelsmannSpringer.

www.vieweg.de

Cover design: Ulrike Weigel, www.CorporateDesignGroup.de
Printing and binding: Druckerei Hubert & Co., Göttingen
Printed on acid-free paper
Printed in Germany

ISBN 3-528-13130-6

TABLE OF CONTENTS

PREFACE

These notes will study the dynamics of iterated holomorphic mappings from a Riemann surface to itself, concentrating on the classical case of rational maps of the Riemann sphere. They are based on introductory lectures given at Stony Brook during the Fall Term of 1989-90 and also in later years. I am grateful to the audiences for a great deal of constructive criticism, and to Branner, Douady, Hubbard, and Shishikura who taught me most of what I know in this field. Also, I want to thank A. Poirier, S. Zakeri, and R. Perez for their extremely helpful criticisms of various drafts.

There have been a number of extremely useful surveys of holomorphic dynamics over the years — those of Brolin, Douady, Blanchard, Lyubich, Devaney, Keen, and Eremenko-Lyubich, as well as the textbooks by Beardon, Steinmetz, and Carleson-Gamelin, are particularly recommended to the reader. (Compare the list of references at the end, and see Alexander for historical information.)

This subject is large and rapidly growing. These lectures are intended to introduce the reader to some key ideas in the field, and to form a basis for further study. The reader is assumed to be familiar with the rudiments of complex variable theory and of two-dimensional differential geometry, as well as some basic topics from topology. The necessary material can be found for example in Ahlfors 1966, Hocking and Young, Munkres, and Willmore. However, two big theorems will be used here without proof, namely the Uniformization Theorem in §1 and the existence of solutions for the measurable Beltrami equation in Appendix F. (See the references in those sections.)

John Milnor, Stony Brook, June 1999

CHRONOLOGICAL TABLE

Following is a list of some of the founders of the field of complex dynamics.

Ernst Schröder	1841-1902
Hermann Amandus Schwarz	1843–1921
Henri Poincaré	1854–1912
Gabriel Kœnigs	1858–1931
Léopold Leau	1868–1940(?)
Lucjan Emil Böttcher	1872– ?
Samuel Lattès	1873-1918
Constantin Carathéodory	1873–1950
Paul Montel	1876–1975
Pierre Fatou	1878–1929
Paul Koebe	1882–1945
Arnaud Denjoy	1884–1974
Gaston Julia	1893–1978
Carl Ludwig Siegel	1896–1981
Hubert Cremer	1897–1983
Herbert Grötzsch	1902–1993
Charles Morrey	1907–1984
Lars Ahlfors	1907–1996
Lipman Bers	1914–1993

Among the many present day workers in the field, let me mention a few whose work is emphasized in these notes: I. Noel Baker (1932), Adrien Douady (1935), Dennis P. Sullivan (1941), Michael R. Herman (1942), Bodil Branner (1943), John Hamal Hubbard (1945), William P. Thurston (1946), Mary Rees (1953), Jean-Christophe Yoccoz (1955), Curtis McMullen (1958), Mikhail Y. Lyubich (1959), and Mitsuhiro Shishikura (1960).

RIEMANN SURFACES

§1. Simply Connected Surfaces

The first three sections will present an overview of some background material.

If $V \subset \mathbb{C}$ is an open set of complex numbers, a function $f : V \to \mathbb{C}$ is called *holomorphic* (or "complex analytic") if the first derivative

$$z \mapsto f'(z) = \lim_{h \to 0} (f(z+h) - f(z))/h$$

is defined and continuous as a function from V to $\mathbb{C}$, or equivalently if f has a power series expansion about any point $z_0 \in V$ which converges to f in some neighborhood of z_0. (See for example Ahlfors 1966.) Such a function is *conformal* if the derivative $f'(z)$ never vanishes, and *univalent* if it is conformal and one-to-one. (In particular, our conformal maps must always preserve orientation.)

By a *Riemann surface* S we mean a connected complex analytic manifold of complex dimension one. Thus S is a connected Hausdorff space. Furthermore, in some neighborhood U of an arbitrary point of S we can choose a *local uniformizing parameter* (or "coordinate chart") which maps U homeomorphically onto an open subset of the complex plane $\mathbb{C}$, with the following property: In the overlap $U \cap U'$ between two such neighborhoods, each of these local uniformizing parameters can be expressed as a holomorphic function of the other.

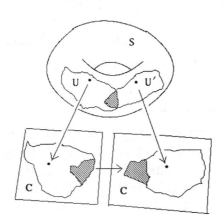

By definition, two Riemann surfaces S and S' are *conformally isomorphic* (or *biholomorphic*) if and only if there is a homeomorphism from S

onto S' which is holomorphic, in terms of the respective local uniformizing parameters. (It is an easy exercise to show that the inverse map $S' \to S$ must then also be holomorphic.) In the special case $S = S'$, such a conformal isomorphism $S \to S$ is called a *conformal automorphism* of S.

Although there is an uncountable infinity of conformally distinct Riemann surfaces, the classification is very easy to describe in the simply connected case. (By definition, the surface S is *simply connected* if every map from a circle into S can be continuously deformed to a constant map. Compare §2.) According to Poincaré and Koebe, there are only three simply connected Riemann surfaces, up to isomorphism:

> **1.1. Uniformization Theorem.** *Any simply connected Riemann surface is conformally isomorphic either*
>
> (a) *to the plane $\mathbb{C}$ consisting of all complex numbers $z = x + iy$,*
>
> (b) *to the open disk $\mathbb{D} \subset \mathbb{C}$ consisting of all z with $|z|^2 = x^2 + y^2 < 1$, or*
>
> (c) *to the Riemann sphere $\hat{\mathbb{C}}$ consisting of $\mathbb{C}$ together with a point at infinity, using $\zeta = 1/z$ as local uniformizing parameter in a neighborhood of the point at infinity.*

This is a generalization of the classical Riemann Mapping Theorem. We will refer to these three cases as the *Euclidean, hyperbolic,* and *spherical* cases respectively. (Compare §2.) I will not try to give a proof of 1.1. The proof, which is quite difficult, may be found in Springer, or Farkas & Kra, or Ahlfors (1973), or Nevanlinna, or in Beardon (1984). (See also Fisher, Hubbard and Wittner.) By assuming this result, we will be able to get more quickly to interesting ideas in holomorphic dynamics. □

The Open Disk $\mathbb{D}$.

For the rest of this section, we will discuss these three surfaces in more detail. We begin with a study of the unit disk $\mathbb{D}$.

> **1.2. Schwarz Lemma.** *If $f : \mathbb{D} \to \mathbb{D}$ is a holomorphic map with $f(0) = 0$, then the derivative at the origin satisfies $|f'(0)| \le 1$. If equality holds, $|f'(0)| = 1$, then f is a rotation about the origin. That is, $f(z) = cz$ for some constant $c = f'(0)$ on the unit circle. On the other hand, if $|f'(0)| < 1$, then $|f(z)| < |z|$ for all $z \ne 0$.*

(The Schwarz Lemma was first proved, in this generality, by Carathéodory.)

Remarks. If $|f'(0)| = 1$, it follows that f is a conformal automorphism of the unit disk. But if $|f'(0)| < 1$ then f cannot be a conformal automorphism of $\mathbb{D}$, since the composition with any $g : (\mathbb{D}, 0) \to (\mathbb{D}, 0)$ would have derivative $g'(0)f'(0) \neq 1$. The example $f(z) = z^2$ shows that f may map $\mathbb{D}$ *onto* itself even when $|f(z)| < |z|$ for all $z \neq 0$ in $\mathbb{D}$.

Proof of 1.2. We use the *Maximum Modulus Principle*, which asserts that a non-constant holomorphic function cannot attain its maximum absolute value at any interior point of its region of definition. First note that the quotient function $q(z) = f(z)/z$ is well defined and holomorphic throughout the disk $\mathbb{D}$ (as one sees by dividing the local power series for f by z). Since $|q(z)| < 1/r$ when $|z| = r < 1$, it follows by the Maximum Modulus Principle that $|q(z)| < 1/r$ for all z in the disk $|z| \leq r$. Since this is true for all $r \to 1$, it follows that $|q(z)| \leq 1$ for all $z \in \mathbb{D}$. Again by the Maximum Modulus Principle, we see that the case $|q(z)| = 1$, for some z in the open disk, can occur only if the function $q(z)$ is constant. If we exclude this case $f(z)/z \equiv c$, then it follows that $|q(z)| = |f(z)/z| < 1$ for all $z \neq 0$, and similarly that $|q(0)| = |f'(0)| < 1$. $\square$

Here is a useful variant statement.

1.2′. Cauchy Derivative Estimate. *If f maps the disk of radius r about z_0 into some disk of radius s, then*

$$|f'(z_0)| \leq s/r .$$

Proof. This follows easily from the Cauchy integral formula (see for example Ahlfors): Set $g(z) = f(z + z_0) + \text{constant}$, so that g maps the disk $\mathbb{D}_r$ centered at the origin to the disk $\mathbb{D}_s$ centered at the origin. Then

$$f'(z_0) = g'(0) = \frac{1}{2\pi i} \oint_{|z|=r_1} \frac{g(z)\, dz}{z^2}$$

for all $r_1 < r$, and the conclusion follows. $\square$

(An alternative proof, based on the Schwarz Lemma, is described in Problem 1-a below. With an extra factor of 2 on the right, this inequality would follow immediately from 1.2 simply by linear changes of variable, since the target disk of radius s must be contained in the disk of radius $2s$ centered at the image $f(z_0)$.)

As an easy corollary, we obtain the following.

1.3. Theorem of Liouville. *A bounded function f which is defined and holomorphic everywhere on $\mathbb{C}$ must be constant.*

For in this case we have s fixed but r arbitrarily large, hence f' must be identically zero. $\square$

As another corollary, we see that our three model surfaces really are distinct. There are natural inclusion maps $\mathbb{D} \to \mathbb{C} \to \widehat{\mathbb{C}}$. Yet it follows from the Maximum Modulus Principle that every holomorphic map $\widehat{\mathbb{C}} \to \mathbb{C}$ must be constant, and from Liouville's Theorem that every holomorphic map $\mathbb{C} \to \mathbb{D}$ must be constant.

Another closely related statement is the following. Let U be an open subset of $\mathbb{C}$.

1.4. Weierstrass Uniform Convergence Theorem. *If a sequence of holomorphic functions $f_n : U \to \mathbb{C}$ converges uniformly to the limit function f, then f itself is holomorphic. Furthermore, the sequence of derivatives f_n' converges, uniformly on any compact subset of U, to the derivative f'.*

It follows that the sequence of second derivatives f_n'' converges, uniformly on compact subsets, to f'', and so on.

Proof of 1.4. Note first that the sequence of first derivatives f_n', restricted to any compact subset $K \subset U$, converges uniformly. For example if $|f_n(z) - f_m(z)| < \epsilon$ for $m, n > N$, and if the r-neighborhood of any point of K is contained in U, then it follows from 1.2' that $|f_n'(z) - f_m'(z)| < \epsilon/r$ for $m, n > N$ and for all $z \in K$. This proves uniform convergence of $\{f_n'\}$ restricted to K to some limit function g, which is necessarily continuous since any uniform limit of continuous functions is continuous. It follows that the integral of f_n' along any path in U converges to the integral of g along this path. Thus $f = \lim f_n$ is an indefinite integral of g, hence g can be identified with the derivative of f. Thus f has a continuous complex first derivative, and hence is a holomorphic function.   □

Conformal Automorphism Groups.

For any Riemann surface S, the notation $\mathcal{G}(S)$ will be used for the group consisting of all conformal automorphisms of S. The identity map will be denoted by $I = I_S \in \mathcal{G}(S)$.

We first consider the case of the Riemann sphere $\widehat{\mathbb{C}}$, and show that $\mathcal{G}(\widehat{\mathbb{C}})$ can be identified with a well known complex Lie group. Thus $\mathcal{G}(\widehat{\mathbb{C}})$ is not only a group, but is also a complex manifold; and the product and inverse operations for this group are both holomorphic maps.

1.5. Lemma. Möbius transformations. *The group $\mathcal{G}(\widehat{\mathbb{C}})$ of all conformal automorphisms of the Riemann sphere is equal to the group of all **fractional linear transformations** (also called **Möbius transformations**)*

$$g(z) = (az + b)/(cz + d),$$

where the coefficients are complex numbers with determinant
$ad - bc \neq 0$.

Here, if we multiply numerator and denominator by a common factor, then it is always possible to normalize so that $ad - bc = 1$. The resulting coefficients are well defined up to a simultaneous change of sign. *Thus it follows that the group $\mathcal{G}(\widehat{\mathbb{C}})$ of conformal automorphisms can be identified with the complex 3-dimensional Lie group $\mathbf{PSL}(2, \mathbb{C})$, consisting of all 2×2 complex matrices with determinant $+1$ modulo the subgroup $\{\pm I\}$.* (Since the complex dimension is 3, it follows that the real dimension of $\mathbf{PSL}(2, \mathbb{C})$ is 6.)

Proof of 1.5. It is easy to check that $\mathcal{G}(\widehat{\mathbb{C}})$ contains this group of fractional linear transformations as a subgroup. After composing the given $g \in \mathcal{G}(\widehat{\mathbb{C}})$ with a suitable element of this subgroup, we may assume that $g(0) = 0$ and $g(\infty) = \infty$. But then the quotient $g(z)/z$ is a bounded holomorphic function from $\mathbb{C} \smallsetminus \{0\}$ to itself. (In fact $g(z)/z$ tends to the non-zero finite value $g'(0)$ as $z \to 0$. Setting $\zeta = 1/z$ and $G(\zeta) = 1/g(1/\zeta)$, evidently $g(z)/z = \zeta/G(\zeta)$ tends to the non-zero finite value $1/G'(0)$ as $z \to \infty$.) Setting $z = e^w$, it follows that the composition $w \mapsto g(e^w)/e^w$ is a bounded holomorphic function on $\mathbb{C}$. Hence it takes a constant value c by Liouville's Theorem. Therefore $g(z) = cz$ is linear, hence g itself is an element of $\mathbf{PSL}(2, \mathbb{C})$. $\square$

Next we will show that both $\mathcal{G}(\mathbb{C})$ and $\mathcal{G}(\mathbb{D})$ can be considered as Lie subgroups of $\mathcal{G}(\widehat{\mathbb{C}})$.

1.6. Corollary. The affine group. *The group $\mathcal{G}(\mathbb{C})$ of all conformal automorphisms of the complex plane consists of all affine transformations*

$$f(z) = \lambda z + c$$

with complex coefficients $\lambda \neq 0$ and c .

Proof. First note that every conformal automorphism f of $\mathbb{C}$ extends uniquely to a conformal automorphism of $\widehat{\mathbb{C}}$. In fact $\lim_{z \to \infty} f(z) = \infty$, so the singularity of $1/f(1/\zeta)$ at $\zeta = 0$ is removable. (Compare Ahlfors, 1966, p.124.) It follows that $\mathcal{G}(\mathbb{C})$ can be identified with the subgroup of $\mathcal{G}(\widehat{\mathbb{C}})$ consisting of Möbius transformations which fix the point ∞ . Evidently this is just the complex 2-dimensional subgroup consisting of all complex affine transformations of $\mathbb{C}$. $\square$

1.7. Theorem. Automorphisms of $\mathbb{D}$. *The group $\mathcal{G}(\mathbb{D})$ of all conformal automorphisms of the unit disk can be identified* *with the subgroup of* $\mathcal{G}(\hat{\mathbb{C}})$ *consisting of all maps*

$$f(z) \;=\; e^{i\theta}\frac{z-a}{1-\bar{a}z} \tag{1:1}$$

where a *ranges over the open disk* $\mathbb{D}$ *and where* $e^{i\theta}$ *ranges over the unit circle* $\partial\mathbb{D}$.

This is no longer a *complex* Lie group. However, $\mathcal{G}(\mathbb{D})$ is a *real* 3-dimensional Lie group, having the topology of a "solid torus" $\mathbb{D}\times\partial\mathbb{D}$.

Proof of 1.7. Evidently the map f defined by (1:1) carries the entire Riemann sphere $\hat{\mathbb{C}}$ conformally onto itself. A brief computation shows that

$$|f(z)| < 1 \quad\Longleftrightarrow\quad (z-a)(\bar{z}-\bar{a}) < (1-\bar{a}z)(1-a\bar{z})$$
$$\Longleftrightarrow\quad (1-z\bar{z})\,(1-a\bar{a}) > 0 \,.$$

For any $a \in \mathbb{D}$, it follows that $|f(z)| < 1 \Longleftrightarrow |z| < 1$. Hence f maps $\mathbb{D}$ onto itself. Now if $g : \mathbb{D} \xrightarrow{\cong} \mathbb{D}$ is an arbitrary conformal automorphism, and $a \in \mathbb{D}$ is the unique solution to the equation $g(a) = 0$, then we can consider $f(z) = (z-a)/(1-\bar{a}z)$, which also maps a to zero. The composition $g \circ f^{-1}$ is an automorphism fixing the origin, hence it has the form $g \circ f^{-1}(z) = e^{i\theta}z$ by the Schwarz Lemma, and $g(z) = e^{i\theta}f(z)$, as required. $\quad\square$

It is often more convenient to work with the *upper half-plane* $\mathbb{H}$, consisting of all complex numbers $w = u + iv$ with $v > 0$.

1.8. Lemma. $\mathbb{D} \cong \mathbb{H}$. *The half-plane* $\mathbb{H}$ *is conformally isomorphic to the disk* $\mathbb{D}$ *under the holomorphic mapping*

$$w \mapsto (i-w)/(i+w) \,,$$

with inverse

$$z \mapsto i(1-z)/(1+z) \,,$$

where $z \in \mathbb{D}$ *and* $w \in \mathbb{H}$.

Proof. If z and $w = u + iv$ are complex numbers related by these formulas, then $|z|^2 < 1$ if and only if $|i-w|^2 = u^2 + (1-2v+v^2)$ is less than $|i+w|^2 = u^2 + (1+2v+v^2)$, or in other words if and only if $v > 0$. $\quad\square$

1.9. Corollary. Automorphisms of $\mathbb{H}$. *The group* $\mathcal{G}(\mathbb{H})$ *consisting of all conformal automorphisms of the upper half-plane can be identified with the group of all fractional linear transformations* $w \mapsto (aw+b)/(cw+d)$ *where the coefficients* a, b, c, d *are real with determinant* $ad - bc > 0$.

If we normalize so that $ad - bc = 1$, then these coefficients are well defined up to a simultaneous change of sign. *Thus $\mathcal{G}(\mathbb{H})$ is isomorphic to the group $\mathbf{PSL}(2, \mathbb{R})$, consisting of all 2×2 real matrices with determinant $+1$ modulo the subgroup $\{\pm I\}$.*

Proof of 1.9. If $f(w) = (aw + b)/(cw + d)$ with real coefficients and non-zero determinant, then it is easy to check that f maps $\mathbb{R} \cup \infty$ homeomorphically onto itself. Note that the image

$$f(i) = (ai + b)(-ci + d)/(c^2 + d^2)$$

lies in the upper half-plane $\mathbb{H}$ if and only if $ad - bc > 0$. It follows easily that this group $\mathbf{PSL}(2, \mathbb{R})$ of positive real fractional linear transformations acts as a group of conformal automorphisms of $\mathbb{H}$. This group acts transitively. In fact the subgroup consisting of all $w \mapsto aw + b$ with $a > 0$ already acts transitively, since such a map carries the point i to a completely arbitrary point $ai + b \in \mathbb{H}$. Furthermore, $\mathbf{PSL}(2, \mathbb{R})$ contains the group of "rotations"

$$g(w) = (w \cos \theta + \sin \theta)/(-w \sin \theta + \cos \theta), \qquad (1\!:\!2)$$

which fix the point $g(i) = i$ with derivative $g'(i) = e^{2i\theta}$. By 1.2 and 1.8, there can be no further automorphisms fixing i, and it follows easily that $\mathbf{PSL}(2, \mathbb{R}) = \mathcal{G}(\mathbb{H})$. $\quad\square$

To conclude this section, we will try to say something more about the structure of these three groups. For any map $f : X \to X$, it will be convenient to use the notation $\mathrm{Fix}(f) \subset X$ for the set of all fixed points $x = f(x)$. If f and g are commuting maps from X to itself, $f \circ g = g \circ f$, note that

$$f\big(\mathrm{Fix}(g)\big) \subset \mathrm{Fix}(g). \qquad (1\!:\!3)$$

For if $x \in \mathrm{Fix}(g)$, then $f(x) = f \circ g(x) = g \circ f(x)$, hence $f(x) \in \mathrm{Fix}(g)$. We first apply these ideas to the group $\mathcal{G}(\mathbb{C})$ of all affine transformations of $\mathbb{C}$.

1.10. Lemma. Commuting elements of $\mathcal{G}(\mathbb{C})$. *Two non-identity affine transformations of $\mathbb{C}$ commute if and only if they have the same fixed point set.*

It follows easily that any $g \neq I$ in the group $\mathcal{G}(\mathbb{C})$ is contained in a unique maximal abelian subgroup consisting of all f with $\mathrm{Fix}(f) = \mathrm{Fix}(g)$, together with the identity element.

Proof of 1.10. Clearly an affine transformation with two fixed points must be the identity map. If g has just one fixed point z_0, then it follows from $(1\!:\!3)$ that any f which commutes with g fixes this same point.

The set of all such f forms a commutative group, consisting of all $f(z) = z_0 + \lambda(z - z_0)$ with $\lambda \neq 0$. Similarly, if $\text{Fix}(g)$ is the empty set, then g is a translation $z \mapsto z + c$, and $f \circ g = g \circ f$ if and only if f is also a translation. $\square$

Now consider the group $\mathcal{G}(\widehat{\mathbb{C}})$ of automorphisms of the Riemann sphere. By definition, an automorphism g is called an *involution* if $g \circ g = I$ but $g \neq I$.

> **1.11. Theorem. Commuting elements of $\mathcal{G}(\widehat{\mathbb{C}})$.** *For every $f \neq I$ in $\mathcal{G}(\widehat{\mathbb{C}})$, the set $\text{Fix}(f) \subset \widehat{\mathbb{C}}$ contains either one point or two points. In general, two non-identity elements $f, g \in \mathcal{G}(\widehat{\mathbb{C}})$ commute if and only if $\text{Fix}(f) = \text{Fix}(g)$. The only exceptions to this statement are provided by pairs of commuting involutions, each of which interchanges the two fixed points of the other.*

(Compare Problem 1-c. As an example, the involution $f(z) = -z$ with $\text{Fix}(f) = \{0, \infty\}$ commutes with the involution $g(z) = 1/z$ with $\text{Fix}(g) = \{\pm 1\}$.)

Proof of 1.11. The fixed points of a fractional linear transformation can be determined by solving a quadratic equation, so it is easy to check that there must be at least one and at most two distinct solutions in the extended plane $\widehat{\mathbb{C}}$. (If an automorphism of $\widehat{\mathbb{C}}$ fixes three distinct points, then it must be the identity map.)

If f commutes with g, which has exactly two fixed points, then since $f(\text{Fix}(g)) = \text{Fix}(g)$ by (1:3), it follows that f must either have the same two fixed points, or must interchange the two fixed points of g. In the first case, taking the fixed points to be 0 and ∞, it follows that both f and g belong to the commutative group consisting of all linear maps $z \mapsto \lambda z$ with $\lambda \in \mathbb{C} \setminus \{0\}$. In the second case, if f interchanges 0 and ∞, then it is necessarily a transformation of the form $f(z) = \eta/z$, with $f \circ f(z) = z$. Setting $g(z) = \lambda z$, the equation $g \circ f = f \circ g$ reduces to $\lambda^2 = 1$, so that g must also be an involution.

Finally, suppose that g has just one fixed point, which we may take to be the point at infinity. Then by (1:3) any f which commutes with g must also fix the point at infinity. Hence we are reduced to the situation of 1.10, and both f and g must be translations $z \mapsto z + c$. (Such automorphisms with just one fixed point, at which the first derivative is necessarily $+1$, are called *parabolic* automorphisms.) This completes the proof. $\square$

We want a corresponding statement for the open disk $\mathbb{D}$. However it is better to work with the closed disk $\overline{\mathbb{D}}$, in order to obtain a richer set of fixed points. Using 1.7, we see easily that every automorphism of the open disk extends uniquely to an automorphism of the closed disk.

1.12. Theorem. Commuting elements of $\mathcal{G}(\overline{\mathbb{D}})$. *For every $f \neq I$ in $\mathcal{G}(\overline{\mathbb{D}})$, the set $\mathrm{Fix}(f) \subset \overline{\mathbb{D}}$ consists either of a single point of the open disk $\mathbb{D}$, a single point of the boundary circle $\partial \mathbb{D}$, or two points of $\partial \mathbb{D}$. Two non-identity automorphisms $f, g \in \mathcal{G}(\overline{\mathbb{D}})$ commute if and only if they have the same fixed point set in $\overline{\mathbb{D}}$.*

Remark 1.13. (Compare Problem 1-d.) An automorphism of $\overline{\mathbb{D}}$ is often described as 'elliptic', 'parabolic', or 'hyperbolic' according as it has one interior fixed point, one boundary fixed point, or two boundary fixed points. We can describe these transformations geometrically as follows. In the elliptic case, after conjugating by a transformation which carries the fixed point to the origin, we may assume that $0 = g(0)$. It then follows from the Schwarz Lemma that g is just a rotation about the origin. In the parabolic case, it is convenient to replace $\mathbb{D}$ by the upper half-plane, choosing the isomorphism $\mathbb{D} \cong \mathbb{H}$ so that the boundary fixed point corresponds to the point at infinity. Using 1.9, we see that g must correspond to a linear transformation $w \mapsto aw + b$ with a, b real and $a > 0$. Since there are no fixed points in $\mathbb{R} \subset \partial \mathbb{H}$, it follows that $a = 1$, so that we have a horizontal translation. Similarly, in the hyperbolic case, taking the fixed points to be $0, \infty \in \partial \mathbb{H}$, we see that g must correspond to a linear map of the form $w \mapsto aw$ with $a > 0$. It is rather inelegant that we must extend to the boundary in order to distinguish between the parabolic and hyperbolic cases. For a more intrinsic interpretation of this dichotomy, see Problem 1-f or 2-e below.

Proof of 1.12. In fact every automorphism of $\mathbb{D}$ or $\overline{\mathbb{D}}$ is a Möbius transformation, and hence extends uniquely to an automorphism F of the entire Riemann sphere. This extension commutes with the inversion map $\alpha(z) = 1/\bar{z}$. In fact the composition $\alpha \circ F \circ \alpha$ is a holomorphic map which coincides with F on the unit circle, and hence coincides with F everywhere. Thus F has a fixed point z in the open disk $\mathbb{D}$, if and only if it has a corresponding fixed point $\alpha(z)$ in the exterior $\widehat{\mathbb{C}} \smallsetminus \overline{\mathbb{D}}$. It now follows from 1.11 that two elements of $\mathcal{G}(\overline{\mathbb{D}})$ commute if and only if they have the same fixed point set in $\overline{\mathbb{D}}$, providing that we can exclude the possibility of two commuting involutions.

If $F \in \mathcal{G}(\widehat{\mathbb{C}})$ is an involution, note that the derivative $F'(z)$ at each of the two fixed points is -1. Thus, if F maps $\mathbb{D}$ onto itself, neither of these fixed points can be on the boundary circle, hence one fixed point must be in $\mathbb{D}$ and one in $\widehat{\mathbb{C}} \smallsetminus \overline{\mathbb{D}}$. Now a second involution which commutes with F must interchange these two fixed points, and hence cannot map $\mathbb{D}$ onto itself. This completes the proof. $\quad\square$

We conclude this section with a number of problems for the reader.

Problem 1-a. Alternate proof of 1.2'. Check that an arbitrary conformal automorphism

$$g(z) = e^{i\theta}(z - a)/(1 - \bar{a}z)$$

of the unit disk satisfies $|g'(0)| = |1 - a\bar{a}| \leq 1$. Since any holomorphic map $f : \mathbb{D} \to \mathbb{D}$ can be written as a composition $g \circ h$ where g is an automorphism mapping 0 to $f(0)$ and where h is a holomorphic map which and fixes the origin. Conclude using 1.2 that $|f'(0)| \leq 1$ even when $f(0) \neq 0$. More generally, if f maps the disk of radius r centered at z into some disk of radius s, show that $|f'(z)| \leq s/r$.

Problem 1-b. Triple transitivity. Show that the action of the group $\mathcal{G}(\widehat{\mathbb{C}})$ on $\widehat{\mathbb{C}}$ is *simply 3-transitive*. That is, there is one and only one automorphism which carries three distinct specified points of $\widehat{\mathbb{C}}$ into three other specified points. Similarly, show that the action of $\mathcal{G}(\mathbb{C})$ on $\mathbb{C}$ is simply 2-transitive. (For corresponding statements for the disk $\mathbb{D}$, see Problem 2-d.)

Problem 1-c. Cross-ratios. Show that the group $\mathcal{G}(\widehat{\mathbb{C}})$ is generated by the subgroup of affine transformations $z \mapsto az + b$ together with the inversion $z \mapsto 1/z$. Given four distinct points z_j in $\widehat{\mathbb{C}}$ show that the *cross-ratio**

$$\chi(z_1, z_2, z_3, z_4) = \frac{(z_3 - z_1)(z_4 - z_2)}{(z_2 - z_1)(z_4 - z_3)} \in \mathbb{C} \smallsetminus \{0, 1\}$$

is invariant under fractional linear transformations. (If one of the z_j is the point at infinity, this definition extends by continuity.) Show that χ is real if and only if the four points lie on a straight line or circle. Given two points $z_1 \neq z_2$ show that there is one and only one involution f with $\mathrm{Fix}(f) = \{z_1, z_2\}$, and show that a second involution g with $\mathrm{Fix}(g) = \{z_1', z_2'\}$ commutes with f if and only if $\chi(z_1, z_2, z_1', z_2') = 1/2$.

---

* Caution: Several variant notations for cross-ratios are in common use. This particular version, characterized by the property that $\chi(0, 1, z, \infty) = z$, is particularly convenient for our purposes. Compare Problem 2-c; and note that $\chi(a, b, c, d) > 1$ whenever a, b, c, d are real with $a < b < c < d$.

Problem 1-d. Conjugacy classes in $\mathcal{G}(\mathbb{H})$ **.** By definition, a conformal automorphism of $\mathbb{D}$ or $\mathbb{H}$ is *elliptic* if it has a fixed point, and otherwise is *parabolic* or *hyperbolic* according as its extension to the boundary circle has one fixed point or two fixed points. Classify conjugacy classes in the groups $\mathcal{G}(\mathbb{H}) \cong \mathbf{PSL}(2, \mathbb{R})$ as follows. Show that every automorphism of $\mathbb{H}$ without fixed point is conjugate to a unique transformation of the form $w \mapsto w + 1$ or $w \mapsto w - 1$ or $w \mapsto \lambda w$ with $\lambda > 1$; and show that the conjugacy class of an automorphism g with fixed point $w_0 \in \mathbb{H}$ is uniquely determined by the derivative $\lambda = g'(w_0)$, where $|\lambda| = 1$. Show also that each non-identity element of $\mathbf{PSL}(2, \mathbb{R})$ belongs to one and only one "one-parameter subgroup", and that each one-parameter subgroup is conjugate to either

$$t \;\mapsto\; \begin{bmatrix} 1 & t \\ 0 & 1 \end{bmatrix} \quad \text{or} \quad \begin{bmatrix} e^t & 0 \\ 0 & e^{-t} \end{bmatrix} \quad \text{or} \quad \begin{bmatrix} \cos t & \sin t \\ -\sin t & \cos t \end{bmatrix}$$

according as its elements are parabolic or hyperbolic or elliptic. Here t ranges over the additive group of real numbers.

Problem 1-e. The Euclidean case. Show that the conjugacy class of a non-identity automorphism $g(z) = \lambda z + c$ in the group $\mathcal{G}(\mathbb{C})$ is uniquely determined by its image under the derivative homomorphism $g \mapsto g' \equiv \lambda \in \mathbb{C} \smallsetminus \{0\}$.

Problem 1-f. Anti-holomorphic involutions. By an *anti-holomorphic* mapping from one Riemann surface to another, we mean a transformation which, in terms of local coordinates z and w, takes the form $z \mapsto w = \eta(\bar{z})$ where η is holomorphic. By an *anti-holomorphic involution* of S we mean an anti-holomorphic map $\alpha : S \to S$ such that $\alpha \circ \alpha$ is the identity map.

If L is a straight line in $\mathbb{C}$, show that there is one and only one anti-holomorphic involution of $\mathbb{C}$ having L as fixed point set, and show that no other fixed point sets can occur. Show that the automorphism group $\mathcal{G}(\mathbb{C})$ acts transitively on the set of straight lines in $\mathbb{C}$. Similarly, if L is *either* a straight line or a circle in $\widehat{\mathbb{C}}$, show that there is one and only one anti-holomorphic involution of $\widehat{\mathbb{C}}$ having L as fixed point set, and show that no other non-vacuous (!) fixed point sets can occur. Show that the automorphism group $\mathcal{G}(\widehat{\mathbb{C}})$ acts transitively on the set of straight lines and circles in $\widehat{\mathbb{C}}$. Finally, for an anti-holomorphic involution of $\mathbb{D}$, show that the fixed point set is either a diameter of $\mathbb{D}$ or a circle arc meeting the boundary $\partial\mathbb{D}$ orthogonally, and show that $\mathcal{G}(\mathbb{D})$ acts transitively on the set of all such diameters and circle arcs. Show that an automorphism of $\mathbb{D}$ without interior fixed point is hyperbolic if and only if it commutes with some anti-holomorphic involution, or if and only if it carries some such diameter or circle arc into itself.

Problem 1-g. Fixed points of Möbius transformations. For a non-identity automorphism $g \in \mathcal{G}(\widehat{\mathbb{C}})$, show that the derivatives $g'(z)$ at the two fixed points are reciprocals, say λ and λ^{-1}, and show that the average $\alpha = (\lambda + \lambda^{-1})/2$ is a complete conjugacy class invariant which can take any value in $\mathbb{C}$. (In the special case of a fixed point at infinity, one must evaluate the derivative using the local uniformizing parameter $\zeta = 1/z$.) Show that $\alpha = 1$ if and only if the two fixed points coincide, and that $-1 \leq \alpha = \cos\theta < 1$ if and only if g is conjugate to a rotation through angle θ.

Problem 1-h. Convergence to zero. If a holomorphic map $f : \mathbb{D} \to \mathbb{D}$ fixes the origin, and is not a rotation, prove that the successive images $f^{\circ n}(z)$ converge to zero for all z in the open disk $\mathbb{D}$, and prove that this convergence is uniform on compact subsets of $\mathbb{D}$. (Here $f^{\circ n}$ stands for the n-fold iterate $f \circ \cdots \circ f$. The example $f(z) = z^2$ shows that convergence need not be uniform on all of $\mathbb{D}$.)

§2. Universal Coverings and the Poincaré Metric

First recall some standard topological constructions. (Compare Munk-res, as well as Appendix E.) A map $p : M \to N$ between connected manifolds is called a *covering map* if every point of N has a connected open neighborhood U within N which is *evenly covered*, that is each component of $p^{-1}(U)$ maps onto U by a homeomorphism. The manifold N is *simply connected* if it has no non-trivial coverings; that is, if every such covering map $M \to N$ is a homeomorphism. (Equivalently, N is simply connected if and only if every map from a circle to N can be continuously deformed to a point.) For any connected manifold N, there exists a covering map $\tilde{N} \to N$ such that $\tilde{N}$ is simply connected. This is called the *universal covering* of N, and is unique up to homeomorphism. By a *deck transformation* associated with a covering map $p : M \to N$ we mean a continuous map $\gamma : M \to M$ which satisfies the identity $\gamma = p \circ \gamma$, so that the diagram

$$
\begin{array}{ccc}
M & \xrightarrow{\gamma} & M \\
{\scriptstyle p} \searrow & & \swarrow {\scriptstyle p} \\
& N &
\end{array}
$$

is commutative. For our purposes, the *fundamental group* $\pi_1(N)$ can be defined as the group Γ consisting of all deck transformations for the universal covering $\tilde{N} \to N$. Note that this universal covering is always a *normal* covering of N. That is, given two points x, $x' \in M = \tilde{N}$ with $p(x) = p(x')$, there exists one and only one deck transformation which maps x to x'. It follows that N can be identified with the quotient $\tilde{N}/\Gamma$ of $\tilde{N}$ by this action of Γ. A given group Γ of homeomorphisms of a connected manifold M gives rise in this way to a normal covering $M \to M/\Gamma$ if and only if

(1) Γ acts *properly discontinuously*; that is, any compact set $K \subset M$ intersects only finitely many of its translates $\gamma(K)$ under the action of Γ, and

(2) Γ acts *freely*; that is, every non-identity element of Γ acts without fixed points on M.

Now let S be a Riemann surface. Then the universal covering manifold $\tilde{S}$ inherits the structure of a Riemann surface, and every deck transformation is a conformal automorphism of $\tilde{S}$. According to the Uniformization Theorem 1.1, since this universal covering surface $\tilde{S}$ is simply connected, it must be conformally isomorphic to one of the three model surfaces. Thus we have the following.

2.1. Uniformization Theorem for arbitrary Riemann surfaces. *Every Riemann surface S is conformally isomorphic to a quotient of the form $\tilde{S}/\Gamma$, where $\tilde{S}$ is a simply connected Riemann surface, which is necessarily isomorphic to either $\mathbb{D}$, $\mathbb{C}$, or $\widehat{\mathbb{C}}$, and where $\Gamma \cong \pi_1(S)$ is a group of conformal automorphisms which acts freely and properly discontinuously on $\tilde{S}$.*

The group $\mathcal{G}(\tilde{S})$ consisting of *all* conformal automorphisms of $\tilde{S}$ has been studied in §1. It is a Lie group, and in particular has a natural topology. Since the action of Γ on $\tilde{S}$ is properly discontinuous, it is not difficult to check that Γ must be a *discrete* subgroup of $\mathcal{G}(\tilde{S})$; that is, there exists a neighborhood of the identity element in $\mathcal{G}(\tilde{S})$ which intersects Γ only in the identity element. (Compare Problem 2-a.)

As a curious consequence, we obtain a remarkable property of complex 1-manifolds, which was first proved by Radó. (Compare Ahlfors & Sario.) By definition, a topological space is σ-*compact* if it can be expressed as a countable union of compact subsets.

2.2. Corollary. σ-compactness. *Every Riemann surface can be expressed as a countable union of compact subsets.*

(It can be shown that a connected manifold is σ-compact if and only if it is paracompact, or metrizable, or has a countable basis for the open subsets. However, in general, manifolds need not satisfy any of these conditions.)

Proof of 2.2. This follows from 2.1, since the corresponding property is clearly true for each of the three simply connected surfaces. □

We can now give a very rough catalogue of all possible Riemann surfaces. The discussion will be divided into two easy cases and one hard case.

Spherical Case. According to 1.12, every conformal automorphism of the Riemann sphere $\widehat{\mathbb{C}}$ has at least one fixed point. Therefore, if $S \cong \widehat{\mathbb{C}}/\Gamma$ is a Riemann surface with universal covering $\tilde{S} \cong \widehat{\mathbb{C}}$, then the group $\Gamma \subset \mathcal{G}(\widehat{\mathbb{C}})$ must be trivial, hence S itself must be conformally isomorphic to $\widehat{\mathbb{C}}$.

Euclidean Case. By 1.6, the group $\mathcal{G}(\mathbb{C})$ of conformal automorphisms of the complex plane consists of all affine transformations $z \mapsto \lambda z + c$ with $\lambda \neq 0$. Every such transformation with $\lambda \neq 1$ has a fixed point. Hence, if $S \cong \mathbb{C}/\Gamma$ is a surface with universal covering $\tilde{S} \cong \mathbb{C}$, then Γ must be a discrete group of translations $z \mapsto z + c$ of the complex plane $\mathbb{C}$. There are three subcases:

If Γ is trivial, then S itself is isomorphic to $\mathbb{C}$.

If Γ has just one generator, then S is isomorphic to the infinite *cylinder* $\mathbb{C}/\mathbb{Z}$, where $\mathbb{Z} \subset \mathbb{C}$ is the additive subgroup of integers. Note that this cylinder is isomorphic to the *punctured plane* $\mathbb{C} \smallsetminus \{0\}$ under the isomorphism

$$z \mapsto \exp(2\pi i z) \in \mathbb{C} \smallsetminus \{0\}.$$

If Γ has two generators, then it can be described as a two-dimensional *lattice* $\Lambda \subset \mathbb{C}$, that is an additive group generated by two complex numbers which are linearly independent over $\mathbb{R}$. (Two generators, such as 1 and $\sqrt{2}$, which are dependent over $\mathbb{R}$ would not generate a discrete group.) The quotient $\mathbb{T} = \mathbb{C}/\Lambda$ is called a *torus*.

In all three subcases, note that our surface inherits a locally Euclidean geometry from the Euclidean metric $|dz|$ on its universal covering surface. As an example, the punctured plane $\mathbb{C} \smallsetminus \{0\}$, consisting of points $\exp(2\pi i z) = w$, has a complete locally Euclidean metric $2\pi |dz| = |dw/w|$. (Such a metric is well defined only up to multiplication by a positive constant, since we could equally well use a coordinate of the form $z' = \lambda z + c$ in the universal covering, with $|dz'| = |\lambda\,dz|$. Compare Corollary 1.6.) It will be convenient to use the term *Euclidean surface* for these Riemann surfaces, which admit a complete locally Euclidean metric. The term "*parabolic surface*" is also commonly used in the literature.

Hyperbolic Case. In all other cases, the universal covering $\tilde{S}$ must be conformally isomorphic to the unit disk. Such Riemann surfaces are said to be *hyperbolic*. *It follows from the discussion above that any Riemann surface which is not a torus and not isomorphic to $\hat{\mathbb{C}}$, $\mathbb{C}$ or $\mathbb{C}/\mathbb{Z}$ must necessarily be hyperbolic.* For example any Riemann surface of genus ≥ 2, or more generally any Riemann surface with non-abelian fundamental group, is hyperbolic. (Compare Problem 2-g).)

Remark. Here the word "hyperbolic" is a reference to hyperbolic Geometry, that is the geometry of Lobachevsky and Bolyai. (Compare Corollary 2.10 below.) Unfortunately the term "hyperbolic" has at least three quite distinct well established meanings in holomorphic dynamics. We may refer to a hyperbolic periodic orbit (with multiplier off the unit circle), or to a hyperbolic map (§19), or as a hyperbolic surface as here. In order to avoid confusion, I will sometimes use the more explicit phrase *conformally hyperbolic* when the word is used with this geometric meaning, and reserve the phrase *dynamically hyperbolic* for the other two meanings.

2.3. Examples: The annulus and the punctured disk. We have seen that all Euclidean Riemann surfaces have abelian fundamental group, either trivial or isomorphic to $\mathbb{Z}$ or $\mathbb{Z} \oplus \mathbb{Z}$. The *punctured disk* $\mathbb{D} \smallsetminus \{0\}$ provides an example of a surface which is conformally hyperbolic, with fundamental group isomorphic to $\mathbb{Z}$. The universal covering of $\mathbb{D} \smallsetminus \{0\}$ can be identified with the left half-plane, mapped onto $\mathbb{D} \smallsetminus \{0\}$ by the exponential map. Similarly, the *annulus*

$$\mathbb{A}_r = \{ z \in \mathbb{C} \, ; \, 1 < |z| < r \}$$

is a hyperbolic surface, since it admits a holomorphic map to the unit disk. (Compare Lemma 2.5 below.) The fundamental group $\pi_1(\mathbb{A}_r)$ is again free cyclic. In fact annuli and the punctured disk are the only hyperbolic surfaces with abelian fundamental group, other than the disk itself. A closely related property is that each of these surfaces has a non-trivial Lie group of automorphisms. (Compare Problem 2-g.)

2.4. Example: The triply punctured sphere. If we remove either one or two points from the Riemann sphere $\widehat{\mathbb{C}}$ then we obtain a Euclidean surface, namely the complex plane $\mathbb{C}$ or the punctured plane $\mathbb{C} \smallsetminus \{0\} \cong \mathbb{C}/\mathbb{Z}$. However, if we remove three distinct points then we obtain a surface $S = \mathbb{C} \smallsetminus \{0, 1\}$ which must be hyperbolic, for example since its fundamental group is non-abelian. (For a more elementary proof that S is not on our list of non-hyperbolic surfaces, note that for any sufficiently large compact set $K \subset S$ the complement $S \smallsetminus K$ has at least three connected components. We say that the thrice punctured sphere has three "*ends*", while any non-hyperbolic surface has at most two ends. An explicit universal covering map $\mathbb{H} \to \mathbb{C} \smallsetminus \{0, 1\}$ is given by the "elliptic modular function". Compare Problem 7-g.)

Closely related is the statement that any Riemann surface can be made hyperbolic by removing at most three points. (In the case of a torus, it suffices to remove just one point, since that will correspond to removing infinitely many points in the universal covering of the torus.)

The inclusions $\mathbb{D} \to \mathbb{C} \to \widehat{\mathbb{C}}$ provide examples of non-constant holomorphic maps from the hyperbolic surface $\mathbb{D}$ to the Euclidean surface $\mathbb{C}$ and then to the Riemann sphere $\widehat{\mathbb{C}}$. However, no maps in the other direction are possible:

2.5. Lemma. Maps between surfaces of different type.
Every holomorphic map from a Euclidean Riemann surface to a hyperbolic one is necessarily constant. Similarly, every holomorphic map from the Riemann sphere to a Euclidean or hyperbolic surface is necessarily constant.

For a holomorphic map $f : S \to S'$ can be lifted to a holomorphic map

$\tilde{f} : \tilde{S} \rightarrow \tilde{S}'$ between universal covering surfaces. (Compare Problem 2-b below.) However, as noted in §1, any holomorphic map $\hat{\mathbb{C}} \rightarrow \mathbb{C}$ or $\mathbb{C} \rightarrow \mathbb{D}$ must be constant, by the Maximum Modulus Principle and by Liouville's Theorem.   □

2.6. Picard's Theorem. *Every holomorphic map* $f : \mathbb{C} \rightarrow \mathbb{C}$ *which omits two different values must necessarily be constant.*

This follows immediately from 2.4 and 2.5. For if f omits two values a, b, then it can be considered as a map from the Euclidean surface $\mathbb{C}$ to the hyperbolic surface $\mathbb{C} \smallsetminus \{a, b\}$.   □

The Poincaré Metric.

Every hyperbolic surface has a preferred Riemannian metric, constructed as follows. We first consider the simply connected case.

2.7. Lemma. The Poincaré metric on $\mathbb{D}$ **.** *There exists one and, up to multiplication by a positive constant, only one Riemannian metric on the disk* $\mathbb{D}$ *which is invariant under every conformal automorphism of* $\mathbb{D}$ *.*

As an immediate corollary, we get exactly the same statement for the upper half plane $\mathbb{H}$, or for any other surface which is conformally isomorphic to $\mathbb{D}$.

Proof of 2.7. Geometrically, we can prove this statement as follows. To define a Riemannian metric on a smooth manifold M , we must assign a length $\|v\|$ to every tangent vector v at every point of M . Consider then a tangent vector v to the open disk $\mathbb{D}$ at some point $z_0 \in \mathbb{D}$. Choose an automorphism $g \in \mathcal{G}(\mathbb{D})$ which maps z_0 to the origin. Then the first derivative of g at z_0 yields a linear map Dg_{z_0} from the tangent space of $\mathbb{D}$ at z_0 onto the tangent space at the origin. We define $\|v\|$ to be *twice* the Euclidean length of the image vector $Dg_{z_0}(v)$. (The factor of two is inserted for convenience; compare formula $(2\!:\!3)$ below.) Since g is unique up to composition with a rotation of the disk, this length $\|v\|$ is well defined, and is clearly invariant under all automorphisms of $\mathbb{D}$. Finally, since the correspondence $v \mapsto \|v\|^2$, for tangent vectors at a specified point of $\mathbb{D}$, is clearly a homogeneous quadratic function, this construction does indeed yield a Riemannian metric.

Alternatively, using classical notations, we can prove 2.7 more explicitly as follows. A *Riemannian metric* on a open subset of $\mathbb{C}$ can be described as an expression of the form

$$ds^2 = g_{11}dx^2 + 2g_{12}dxdy + g_{22}dy^2 ,$$

where $[g_{jk}]$ is a positive definite matrix which depends smoothly on the point $z = x + iy$. Such a metric is said to be *conformal* if $g_{11} = g_{22}$ and $g_{12} = 0$, so that the matrix $[g_{ij}]$, evaluated at any point z, is some positive multiple of the identity matrix. In other words, a conformal metric is one which can be written as $ds^2 = \gamma(x+iy)^2(dx^2 + dy^2)$, or briefly as $ds = \gamma(z)|dz|$, where the function $\gamma(z)$ is smooth and strictly positive. By definition, such a metric is *invariant* under a conformal automorphism $w = f(z)$ if and only if it satisfies the identity $\gamma(w)|dw| = \gamma(z)|dz|$, or in other words.

$$\gamma(f(z)) = \gamma(z)/|f'(z)|. \qquad (2:1)$$

Equivalently, an f satisfying this condition is called an *isometry* with respect to the metric.

As an example, suppose that a conformal metric $\gamma(w)|dw|$ on the upper half-plane is invariant under every linear automorphism $f(w) = aw + b$ (where $a > 0$). Since $f(i) = ai + b$, equation (2:1) takes the form $\gamma(ai+b) = \gamma(i)/a$. After multiplying the metric by a positive constant, we may assume that $\gamma(i) = 1$. Thus we are led to the formula $\gamma(u + iv) = 1/v$, or in other words

$$ds = |dw|/v \qquad \text{for} \quad w = u + iv \in \mathbb{H}. \qquad (2:2)$$

In fact, the metric defined by this formula is invariant under *every* conformal automorphism g of $\mathbb{H}$. For, if we select some arbitrary point $w_1 \in \mathbb{H}$ and set $g(w_1) = w_2$, then g can be expressed as the composition of a linear automorphism of the form $g_1(w) = aw + b$ which maps w_1 to w_2 and an automorphism g_2 which fixes w_2. We have constructed the metric (2:2) so that g_1 is an isometry, and it follows from 1.2 and 1.8 that $|g_2'(w_2)| = 1$, so that g_2 is an isometry at w_2. Thus our metric is invariant at an arbitrarily chosen point under an arbitrary automorphism.

To complete the proof of 2.7, we must show that a metric which is invariant under all automorphisms of $\mathbb{D}$ or $\mathbb{H}$ is necessarily conformal. For this purpose, given any point $w_0 \in \mathbb{H}$ choose the unique automorphism f which fixes the point w_0 and has derivative $f'(w_0) = \sqrt{-1}$. A brief computation shows that the induced map on Riemannian metrics takes the expression $g_{11}du^2 + 2g_{12}du\,dv + g_{22}dv^2$ at the point w_0 to the expression $g_{22}dv^2 - 2g_{12}du\,dv + g_{11}dv^2$ at w_0. Thus invariance implies that $g_{11} = g_{22}$ and $g_{12} = 0$ at the arbitrary point w_0, as required. $\square$

Definition. This metric $ds = |dw|/v$ is called the *Poincaré metric* on the upper half-plane $\mathbb{H}$. The corresponding expression on the unit disk $\mathbb{D}$ is

$$ds = 2|dz|/(1 - |z|^2) \qquad \text{for} \quad z \in \mathbb{D}, \qquad (2:3)$$

as can be verified by a brief computation using 1.8 and $(2\!:\!1)$.

Remark. The most basic invariant for a Riemannian metric on a surface S is the *Gaussian curvature* function $K : S \to \mathbb{R}$. Since there is an isometry carrying any point of $\mathbb{D}$ (or of $\mathbb{H}$) to any other point, it follows that the Poincaré metric has *constant* Gaussian curvature. In fact this metric, as defined above, has Gaussian curvature $K \equiv -1$. (Compare Problem 2-h.)

Caution: Some authors call $|dz|/(1 - |z|^2)$ the Poincaré metric on $\mathbb{D}$, and correspondingly call $\frac{1}{2}|dw|/v$ the Poincaré metric on $\mathbb{H}$. These modified metrics have constant Gaussian curvature equal to -4.

Thus there is a preferred Riemannian metric ds on $\mathbb{D}$ or on $\mathbb{H}$. More generally, if S is any hyperbolic surface, then the universal covering $\tilde{S}$ is conformally isomorphic to $\mathbb{D}$, and hence has a preferred metric, which is invariant under all conformal isomorphisms of $\tilde{S}$. In particular, it is invariant under deck transformations. It follows that there is one and only one Riemannian metric on S so that the projection $\tilde{S} \to S$ is a *local isometry*, mapping any sufficiently small open subset of $\tilde{S}$ isometrically onto its image in S. By definition, the metric ds constructed in this way is called the *Poincaré metric* on the hyperbolic surface S.

2.8. Example. The punctured disk, The universal covering surface for the punctured disk $\mathbb{D} \smallsetminus \{0\}$ can be identified with the left half plane $\{w = u + iv\,;\, u < 0\}$ under the exponential map $w \mapsto z = e^w \in \mathbb{D} \smallsetminus \{0\}$, with $dz/z = dw$. Evidently the Poincaré metric $|dw/u|$ on the left half plane corresponds to the metric $|dz/r \log r|$ on the punctured disk, where $r = |z|$ and $u = \log r$. (Thus the circle $|z| = r$ has length $2\pi/|\log r|$, which tends to zero as $r \to 0$, although this circle has infinite Poincaré distance from the boundary point $z = 0$.) A neighborhood of zero, intersected with $\mathbb{D} \smallsetminus \{0\}$ can be embedded isometrically as a surface of revolution in Euclidean 3-space. (The generating curve is known as a "tractrix".)

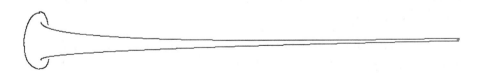

Definition. Let S be a hyperbolic surface with Poincaré metric ds. The integral $\int_P ds$ along any piecewise smooth path $P : [0,1] \to S$ is called the *Poincaré length* of this path. For any two points z_1 and z_2 in S, the *Poincaré distance* $\mathrm{dist}(z_1\,,\,z_2) = \mathrm{dist}_S(z_1\,,\,z_2)$ is defined to be the infimum, over all piecewise smooth paths P joining z_1 to z_2, of the Poincaré length $\int_P ds$. In fact we will see that there always exist a path of minimal length:

2.9. Completeness Lemma. *Every hyperbolic surface S is complete with respect to its Poincaré metric. That is:*

(a) *every Cauchy sequence with respect to the metric* dist_S *converges, or equivalently:*

(b) *every closed neighborhood*

$$N_r(z_0, \mathrm{dist}_S) = \{z \in S \; ; \; \mathrm{dist}_S(z, z_0) \le r\}$$

is a compact subset of S. Furthermore:

(c) *any two points of S are joined by at least one minimal geodesic.*

(In the simply connected case, there is exactly one geodesic between any two points.)

Proof of 2.9. First consider the special case $S = \mathbb{D}$. Given any two points of $\mathbb{D}$ we can first choose a conformal automorphism which moves the first point to the origin and the second to some point r on the positive real axis. For any path P between 0 and r within $\mathbb{D}$ we have

$$\int_P ds = \int_P \frac{2|dz|}{1 - |z|^2} \ge \int_P \frac{2|dx|}{1 - x^2} \ge \int_0^r \frac{2dx}{1 - x^2} = \log\frac{1+r}{1-r},$$

with equality if and only if P is the straight line segment $[0, r]$. For any $z \in \mathbb{D}$, it follows that the Poincaré distance from 0 to z is given by

$$\delta = \mathrm{dist}_\mathbb{D}(0, z) = \log\frac{1 + |z|}{1 - |z|}.$$

(Compare Problem 2-c. Equivalently, we can write $|z| = \tanh(\delta/2)$.) Furthermore, the straight line segment from 0 to z is the unique minimal Poincaré geodesic. This proves (b) and (c) in the simply connected case. The general case follows immediately, and assertion (a) then follows easily. (Compare Willmore.) □

2.10. Corollary. Constant curvature metrics. *Every Riemann surface admits a complete conformal metric with constant curvature which is either positive, negative or zero according as the surface is spherical, hyperbolic, or Euclidean.*

In fact, in the hyperbolic case there is one and only one conformal metric which is complete, with constant Gaussian curvature equal to -1. (Compare Problem 2-i.) In the Euclidean case, the corresponding metric is unique only up to multiplication by a positive constant. In the spherical case, identifying the Riemann sphere $\widehat{\mathbb{C}}$ with the unit sphere in $\mathbb{R}^3$ under stereographic projection, we obtain the standard *spherical metric*

$$ds = 2|dz|/(1+|z|^2), \tag{2:4}$$

with constant Gaussian curvature $+1$. (Problem 2-h.) This spherical metric is smooth and well behaved, even in a neighborhood of the point at infinity. In fact the inversion map $z \mapsto 1/z$ is an isometry. However, the spherical metric is far from unique, since it is not preserved by most Möbius transformations. Its group of orientation preserving isometries $\mathbf{SO}(3)$ is much smaller than the full group $\mathcal{G}(\widehat{\mathbb{C}})$ of all conformal automorphisms. □

Remark. For computer calculations, a more convenient metric for $\widehat{\mathbb{C}}$ is given by the chordal distance formula

$$\text{dist}'(z_1, z_2) = \frac{2|z_1 - z_2|}{\sqrt{(1+|z_1|^2)(1+|z_2|^2)}} = 2\sin(s/2), \tag{2:5}$$

where $s = \text{dist}(z_1, z_2)$ is the usual spherical distance. As an example, using (2:5) the distance between z and the "antipodal" point $-1/\bar{z}$ is always equal to $+2$.

These non-hyperbolic metrics of curvature ≥ 0 are certainly of interest. However, in the hyperbolic case, the Poincaré metric with curvature -1 is of fundamental importance because of its marvelous property of never increasing under holomorphic maps.

2.11. Theorem of Pick. *If $f : S \to S'$ is a holomorphic map between hyperbolic surfaces, then exactly one of the following three statements is valid:*

• *f is a conformal isomorphism from S onto S', and maps S with its Poincaré metric isometrically onto S' with its Poincaré metric.*

• *f is a covering map but is not one-to-one. In this case, it is locally but not globally a Poincaré isometry. Every smooth path $P : [0,1] \to S$ of arclength ℓ in S maps to a smooth path $f \circ P$ of the same length ℓ in S', and it follows that*

$$\text{dist}_{S'}(f(p), f(q)) \leq \text{dist}_S(p, q)$$

for every $p, q \in S$. Here equality holds if p is sufficently close to q, but strict inequality will hold, for example, whenever $f(p) = f(q)$ with $p \neq q$.

• *In all other cases, f strictly decreases all non-zero distances. In fact, for any compact set $K \subset S$ there is a constant $c_K < 1$ so that*

$$\text{dist}_{S'}(f(p), f(q)) \leq c_K \text{ dist}_S(p, q)$$

for every $p, q \in K$, and so that every smooth path in K with arclength ℓ (using the Poincaré metric for S) maps to a path of Poincaré arclength $\leq c_K \ell$ in S'.

Here is an example to illustrate 2.11. The map $f(z) = z^2$ on the disk $\mathbb{D}$ is certainly not a covering map or a conformal automorphism. Hence it is distance decreasing for the Poincaré metric on $\mathbb{D}$. On the other hand, we can also consider f as a map from the *punctured disk* $\mathbb{D} \setminus \{0\}$ to itself. In this case, f is a two-to-one covering map. Hence f is a local isometry for the Poincaré metric on $\mathbb{D} \setminus \{0\}$. In fact, the universal covering of $\mathbb{D} \setminus \{0\}$ can be identified with the left half-plane, mapped onto $\mathbb{D} \setminus \{0\}$ by the exponential map. (Compare 2.8.) Then f lifts to the automorphism $F : w \mapsto 2w$ of this half-plane, which evidently preserves the Poincaré metric.

Proof of 2.11. Let TS_p be the tangent space of S at p. This is a complex one-dimensional vector space. We will think of the Poincaré metric on S as specifying a *norm* $\|v\|$ for each vector $v \in TS_p$, with $\|v\| > 0$ for $v \neq 0$. The holomorphic map $f : S \to S'$ induces a linear first derivative map $Df_p : TS_p \to TS'_{f(p)}$. Let us compare the Poincaré norm $\|v\|$ of a vector $v \in TS_p$ with the Poincaré norm of its image in $TS'_{f(p)}$. Evidently the ratio

$$\|Df_p(v)\| / \|v\|$$

is independent of the choice of non-zero vector v, and can be described as the *norm* $\|Df_p\|$ of the first derivative at p. In the special case of a fixed point $z = f(z)$ of a map on a hyperbolic open subset of $\mathbb{C}$, note that $\|Df_z\|$ can be identified with the absolute value of the classical first derivative $f'(z) = df/dz$. Therefore, for a holomorphic map $f : \mathbb{D} \to \mathbb{D}$ with $f(0) = 0$, the Schwarz Lemma asserts that $\|Df_0\| \leq 1$, with equality if and only if f is a conformal automorphism. More generally, if $f : S \to S'$ is a holomorphic map between simply connected hyperbolic surfaces, and if $p \in S$, it follows immediately that $\|Df_p\| \leq 1$, with equality if and only if f is a conformal isomorphism. Now consider the case where S and S' are not necessarily simply connected. Choose some lifting $F : \tilde{S} \to \tilde{S}'$ to the universal covering surfaces, and some point $\tilde{p}$ over p. From the commutative diagram

$$\begin{array}{ccc} T\tilde{S}_{\tilde{p}} & \longrightarrow & T\tilde{S}'_{F(\tilde{p})} \\ \downarrow & & \downarrow \\ TS_p & \longrightarrow & TS'_{f(p)} \end{array}$$

where the vertical arrows preserve the Poincaré norm, and where both $\tilde{S}$ and $\tilde{S}'$ are conformally isomorphic to $\mathbb{D}$, we see that $\|Df_p\| \leq 1$, with equality if and only if F is a conformal isomorphism from $\tilde{S}$ onto $\tilde{S}'$, or in other words if and only if $f : S \to S'$ is a covering map. (Compare Problem 2-b.)

In particular, if f is not a covering map, then F cannot be a conformal isomorphism, hence $\|Df_p\| < 1$ for all $p \in S$. If K is a compact subset of S, it follows by continuity that $\|Df_p\|$ attains some maximum value $c < 1$ as p varies over K. Now for any smooth path $P : [0,1] \to S$ the derivative DP_t carries the unit tangent vector at $t \in [0,1]$ to a vector in $TS_{P(t)}$ which is called the *velocity vector* $P'(t)$ for the path P at $P(t)$. By definition the Poincaré arclength of P is the integral

$$\mathrm{length}_S(P) = \int_0^1 \|P'(t)\| \, dt .$$

Similarly

$$\mathrm{length}_{S'}(f \circ P) = \int_0^1 \|Df_{P(t)}(P'(t))\| \, dt ,$$

so if $\|Df_p\| \leq c$ throughout K, it follows that

$$\mathrm{length}_{S'}(f \circ P) \leq c \, \mathrm{length}_S(P)$$

for every smooth path within K. In order to compare distances within K, it is necessary to choose some larger compact set $K' \subset S$ so that any two points p and q of K can be joined by a geodesic of length $\mathrm{dist}_S(p,q)$ within K'. If $c_K < 1$ is the maximum of $\|Df_p\|$ for $p \in K'$, then it follows that

$$\mathrm{dist}_{S'}(f(p), f(q)) \leq c_K \, \mathrm{dist}_S(p,q) ,$$

as required. □

Remark. In the distance reducing case, it may happen that there is a uniform constant $c < 1$ so that

$$\mathrm{dist}_{S'}(f(p), f(q)) \leq c \, \mathrm{dist}_S(p,q)$$

for *all* p and q in S. In the special case of a map from S to itself, a standard argument then shows that f has a (necessarily unique) fixed point. (Problem 2-j.) However the example of the map $f(w) = w + i$ from the upper half-plane into itself shows that a distance reducing map need not have a fixed point. Even if f does have a fixed point, it does not follow that such a constant $c < 1$ exists. For example if $f(z) = z^2$, mapping the unit disk onto itself, then a brief computation shows that $\|Df_z\| = 2|z|/(1 + |z|^2)$, taking values arbitrarily close to $+1$.

One important application of 2.11 is to the inclusion $i : S \to S'$ where S' is a hyperbolic Riemann surface and S is a connected open subset. If $S \neq S'$ then it follows from 2.11 that

$$\text{dist}_{S'}(p, q) \ < \ \text{dist}_S(p, q) \tag{2:6}$$

for every $p \neq q$ in S. *Thus distances measured relative to a larger Riemann surface are always smaller.* For sharper forms of this inequality see 3.4, as well as A.8 in the appendix.

Concluding Problems.

Problem 2-a. Properly discontinuous groups. Let S be a simply connected Riemann surface, and let $\Gamma \subset \mathcal{G}(S)$ be a discrete group of automorphisms; that is, suppose that the identity element is an isolated point of Γ within the Lie group $\mathcal{G}(S)$. If every non-identity element of Γ acts on S without fixed points, show that the action of Γ is *properly discontinuous.* That is, for every compact $K \subset S$ show that only finitely many group elements γ satisfy $K \cap \gamma(K) \neq \emptyset$. Show that each $z \in S$ has a neighborhood U whose translates $\gamma(U)$ are pairwise disjoint. Conclude that S/Γ is a well defined Riemann surface with S as its universal covering. (More generally, analogous statements are true for any discrete group of isometries of a Riemannian manifold.) On the other hand, show that the free cyclic group consisting of all transformations $z \mapsto 2^n z$ of $\mathbb{C}$, with $n \in \mathbb{Z}$, forms a discrete subgroup of $\mathcal{G}(\mathbb{C})$, but is not properly discontinuous.

Problem 2-b. Lifting to the universal covering. If $S \cong \mathbb{D}/\Gamma$ and $S' \cong \mathbb{D}/\Gamma'$ are hyperbolic surfaces, show that any holomorphic $f : S \to S'$ lifts to a holomorphic map $\tilde{f} : \mathbb{D} \to \mathbb{D}$, unique up to composition with an element of Γ'. Show that $\tilde{f}$ induces a group homomorphism $\gamma \mapsto \gamma'$ from Γ to Γ' satisfying the identity

$$\tilde{f} \circ \gamma = \gamma' \circ \tilde{f}$$

for every $\gamma \in \Gamma$. Show that f is a covering map if and only if $\tilde{f}$ is a conformal automorphism.

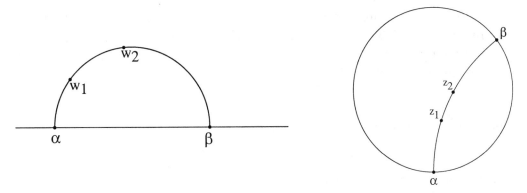

Problem 2-c. Poincaré geodesics. Show that each geodesic for the Poincaré metric in the upper half-plane is a straight line or semi-circle which meets the real axis orthogonally. (Compare Problem 1-f. A related statement, for any Riemannian manifold, is that a curve which is the fixed point set of an isometric involution must necessarily be a geodesic.) If the geodesic through w_1 and w_2 meets $\partial\mathbb{H} = \mathbb{R}\cup\infty$ at the points α and β, show that the Poincaré distance between w_1 and w_2 is equal to the logarithm of the cross-ratio $\chi(\alpha, w_1, w_2, \beta)$, as defined in Problem 1-c. Show that each Poincaré neighborhood $N_r(w_0, \mathrm{dist}_{\mathbb{H}})$ in the upper half-plane is bounded by a Euclidean circle, but that w_0 is not its Euclidean center. Prove corresponding statements for the unit disk.

Problem 2-d. The action of $\mathcal{G}(\mathbb{D})$. Show that the action of the automorphism group $\mathcal{G}(\mathbb{D})$ carries two points of $\mathbb{D}$ into two other specified points if and only if they have the same Poincaré distance. Show that the action of $\mathcal{G}(\mathbb{D})$ on the boundary circle $\partial\mathbb{D}$ carries three specified points into three other specified points if and only if they have the same cyclic order.

Problem 2-e. Classifying automorphisms of $\mathbb{D}$. Show that an automorphism of $\mathbb{H}$ or $\mathbb{D}$ is hyperbolic (Problem 1-d) if and only if it carries some Poincaré geodesic into itself without fixed points.

Problem 2-f. Infinite band, cylinder, and annulus. Define the *infinite band* $B \subset \mathbb{C}$ of height π to be the set of all $z = x + iy$ with $|y| < \pi/2$. Show that the exponential map carries B isomorphically onto the right half-plane. Show that the Poincaré metric on B takes the form

$$ds = |dz|/\cos y. \qquad (2:7)$$

Show that the real axis is a geodesic whose Poincaré arclength coincides with its usual Euclidean arclength, and show that each real translation $z \mapsto z + c$ is a hyperbolic automorphism of B having the real axis as its unique invariant geodesic. For any $c > 0$, form the quotient cylinder

$S_c = B/(c\mathbb{Z})$ by identifying each $z \in B$ with $z + c$. By definition, the *modulus* $\mathrm{mod}(S_c)$ of the resulting cylinder is the ratio π/c of height to circumference. Show that this cylinder, with it Poincaré metric, has a unique simple closed geodesic, with

$$\text{length} \;=\; c \;=\; \pi/\mathrm{mod}(S_c) \;.$$

Show that S_c is conformally isomorphic to the annulus

$$\mathbb{A}_r \;=\; \{\, z \in \mathbb{C} \;;\; 1 < |z| < r \,\}$$

where $\log r = 2\pi^2/c$. Conclude that

$$\mathrm{mod}(\mathbb{A}_r) \;=\; \frac{\log r}{2\pi}$$

is a conformal invariant.

Problem 2-g. Abelian fundamental groups. Show that every hyperbolic surface with abelian fundamental group is conformally isomorphic either to the disk $\mathbb{D}$, or the punctured disk $\mathbb{D} \smallsetminus \{0\}$, or to the annulus $\mathbb{A}_r$ for some uniquely defined $r > 1$. (Compare Theorem 1.12 and Problems 1-e, 2-f.) Show that this annulus has a unique simple closed geodesic, which has length $\ell = 2\pi^2/\log r$. On the other hand, show that the punctured disk $\mathbb{D} \smallsetminus \{0\}$ has no closed geodesic. (Either the punctured disk or the punctured plane $\mathbb{C} \smallsetminus \{0\}$ might reasonably be considered as the limiting case of an annulus, as the modulus tends to infinity.) Show that the conformal automorphism group $\mathcal{G}(\mathbb{D} \smallsetminus \{0\})$ of a punctured disk is isomorphic to the circle group $\mathbf{SO}(2)$, while the conformal automorphism group of an annulus is isomorphic to the non-abelian group $\mathbf{O}(2)$. What is the automorphism group for $\mathbb{C} \smallsetminus \{0\}$? Using Theorems 1.10, 1.12 and Problem 2-b, show that a Riemann surface admits a one-parameter group of conformal automorphisms if and only if its fundamental group is abelian.

Problem 2-h. Gaussian curvature. The Gaussian curvature of a conformal metric $ds = \gamma(w)|dw|$ with $w = u + iv$ is given by the formula

$$K \;=\; \frac{\gamma_u^2 + \gamma_v^2 - \gamma(\gamma_{uu} + \gamma_{vv})}{\gamma^4}$$

where the subscripts stand for partial derivatives. (Compare Willmore, p. 79.) Check that the Poincaré metrics $(2\!:\!2)$, $(2\!:\!3)$ and $(2\!:\!7)$ have curvature $K \equiv -1$, and that the spherical metric $(2\!:\!4)$ has curvature $K \equiv +1$.

Problem 2-i. Metrics of constant curvature. A theorem of Heinz Hopf asserts that for each real number K there is one and only one complete simply connected surface of constant curvature K, up to isometry.

(See Willmore p. 162.) Using this result, show that any non-spherical Riemann surface has one and up to a multiplicative constant only one conformal Riemann metric which is complete, with constant Gaussian curvature. On the other hand, show that the Riemann sphere $\widehat{\mathbb{C}}$ has a 3-dimensional family of distinct conformal metrics with curvature $+1$.

Problem 2-j. Fixed points and contracting maps. If S is hyperbolic, show that a holomorphic map $f : S \to S$ can have at most one fixed point, unless some iterate $f^{\circ k}$ is the identity map. (The case of a covering map from S to itself requires special care.) On the other hand, show that any non-hyperbolic surface has a non-identity holomorphic map with more than one fixed point.

We will say that a map $f : X \to X$ from a metric space to itself is *strictly contracting* if there is a constant $0 < c < 1$ so that

$$\text{dist}(f(x), f(y)) \leq c \, \text{dist}(x, y) \tag{2:8}$$

for every $x, y \in X$. If X is a *complete* metric space, show that all orbits under a strictly contracting map must converge towards a unique fixed point. In particular, this statement applies to a self-map of a hyperbolic surface, whenever $(2:8)$ is satisfied. (However, the example $z \mapsto z^2$ on the unit disk shows that a map with a unique fixed point need not satisfy $(2:8)$; and the example $w \mapsto w + i$ on the upper half-plane shows that a map which simply reduces Poincaré distance need not have any fixed point.)

Problem 2-k. No non-trivial holomorphic attractors. In real dynamics, one often encounters extremely complicated *attractors*, that is, compact sets K with $f(K) = K$ such that for any orbit $x_0 \mapsto x_1 \mapsto x_2 \mapsto \cdots$ in some neighborhood of K the distance $\text{dist}(x_n, K)$ converges uniformly to zero. Show that no such behavior can occur for a holomorphic $f : S \to S$. If $K \subset S$ is compact with $f(K) = K$, and if f maps some connected hyperbolic neighborhood U of K into a proper subset of itself, show that f must be strictly contracting on K with respect to the metric dist_U, and hence that K must consist of a single point.

Problem 2-l. The Picard Theorem near infinity. For any holomorphic map $f : \mathbb{D} \smallsetminus \{0\} \to \mathbb{D}$, use the Poincaré metric to show that the image of a small annulus $\epsilon_1 \leq |z| \leq \epsilon_2$ is a region bounded by two small loops, and hence that it is also small. Conclude that f extends continuously, and hence holomorphically,[*] to a map $\mathbb{D} \to \mathbb{D}$. Similarly, show that any holomorphic map from $\mathbb{D} \smallsetminus \{0\}$ to a triply punctured sphere $\widehat{\mathbb{C}} \smallsetminus \{a, b, c\}$ extends to a holomorphic map $\mathbb{D} \to \widehat{\mathbb{C}}$. Use this result to

---

[*] See for example Ahlfors (1966), p. 124.

prove the strong Picard Theorem:

> **Assertion.** *If $f : \mathbb{C} \to \mathbb{C}$ is holomorphic, but not a polynomial, then for every neighborhood $\mathbb{C} \smallsetminus \overline{\mathbb{D}}_r$ of infinity the image $f(\mathbb{C} \smallsetminus \overline{\mathbb{D}}_r)$ omits at most a single point of $\mathbb{C}$. In fact, f takes on every value in $\mathbb{C}$, with at most one exception, infinitely often.*

§3. Normal Families: Montel's Theorem

Let S and T be Riemann surfaces. This section will study compactness in the function space $\mathrm{Hol}(S,T)$ consisting of all holomorphic maps with source S and target T. We first define a topology on this space, and on the larger space $\mathrm{Map}(S,T)$ consisting of all continuous maps from S to T. This topology is known to complex analysts as the *topology of uniform convergence on compact subsets*, or more succinctly as the *topology of locally uniform convergence*. It is known to topologists as the *compact-open topology* (Problem 3-a), or when dealing with smooth manifolds as the C^0-*topology*.

Definition. Let X be a locally compact space and let Y be a metric space. For any f in the space $\mathrm{Map}(X,Y)$ of continuous maps from X to Y we define a family $N_{K,\epsilon}(f)$ of *basic neighborhoods* of f as follows. For any compact subset $K \subset X$ and any $\epsilon > 0$, let $N_{K,\epsilon}(f)$ be the set of all $g \in \mathrm{Map}(X,Y)$ satisfying the condition that

$$\mathrm{dist}(f(x), g(x)) < \epsilon \qquad \text{for all} \qquad x \in K.$$

A subset $\mathcal{U} \subset \mathrm{Map}(X,Y)$ is defined to be *open* if and only if, for every $f \in \mathcal{U}$, there exist K and ϵ as above so that the basic neighborhood $N_{K,\epsilon}(f)$ is contained in $\mathcal{U}$.

> **Lemma 3.1. The topology of locally uniform convergence.** *With these definitions, $\mathrm{Map}(X,Y)$ is a well defined Hausdorff space. A sequence of maps $f_i \in \mathrm{Map}(X,Y)$ converges to the limit g in this topology if and only if*
>
> > (1) *for every compact $K \subset X$, the sequence of maps $f_i|_K : K \to Y$ converges uniformly to $g|_K$,*
>
> *or equivalently if and only if*
>
> > (2) *every point of X has a neighborhood N so that the sequence $\{f_i|_N\}$ converges uniformly to $g|_N$.*
>
> *This topology on $\mathrm{Map}(X,Y)$ depends only on the topologies of X and Y, and not on the particular choice of metric for Y. Furthermore, if X is σ-compact, then $\mathrm{Map}(X,Y)$ is itself a metrizable topological space.*

Proof. The first two statements follow immediately from the definitions. To prove that the topology is independent of the metric on Y, we describe a slightly different form of the definition, which depends only on the topology of Y. Let U be any neighborhood of the diagonal in the product space $Y \times Y$. For any compact $K \subset X$ and any $f \in \mathrm{Map}(X,Y)$, let

$$N_{K,U}(f) \;=\; \{\, g \in \mathrm{Map}(X,Y) \;;\; (f(x), g(x)) \in U \quad \text{for all} \quad x \in K \,\} \,.$$

Given K and U, it is not difficult to construct an $\epsilon > 0$ so that every pair $(f(x), y)$ with $x \in K$ and $\mathrm{dist}(f(x), y) < \epsilon$ belongs to this set U, and it then follows that $N_{K,\epsilon}(f) \subset N_{K,U}(f)$. On the other hand, if $U(\epsilon)$ is the set of all pairs (y, y') with $\mathrm{dist}(y, y') < \epsilon$, then $N_{K,\epsilon}(f) = N_{K,U(\epsilon)}(f)$. Thus, if we take $\{N_{K,U}(f)\}$ as our collection of "basic neighborhoods", then we obtain the same topology, without mentioning any particular choice of metric.

Now suppose that X is σ-compact, that is, suppose that X is a countable union of compact subsets. Since X is also assumed to be locally compact, we can choose compact sets $K_1 \subset K_2 \subset \ldots$ with union X, so that each K_n is contained in the interior of K_{n+1}. It will be convenient to replace the given metric $\mathrm{dist}(y, y')$ by the bounded metric

$$\mu(y, y') \;=\; \mathrm{Min}(\,\mathrm{dist}(y, y')\,,\, 1\,) \;\leq\; 1 \,,$$

which evidently gives rise to the same topology. Define the "locally uniform distance" between two maps from X to Y by the formula

$$\mu'(f, g) \;=\; \sum_n \frac{1}{2^n} \, \mathrm{Max}\{\, \mu(f(x), g(x)) \;;\; x \in K_n \} \,.$$

We must show that this metric gives rise to the required topology on $\mathrm{Map}(X,Y)$. Let $N'_\epsilon(f)$ be the ϵ-neighborhood of f in this metric μ'. Given ϵ, we can choose n so that $1/2^n < \epsilon/2$, and set $K = K_n$. It is then easy to check that $N_{K,\epsilon/2}(f) \subset N'_\epsilon(f)$. Conversely, given K and ϵ we can choose n so that $K \subset K_n$ and check that $N'_{\epsilon/2^n}(f) \subset N_{K,\epsilon}(f)$. Thus the two topologies are indeed the same. $\square$

Now we can specialize to maps between two Riemann surfaces S and T. Since every Riemann surface is σ-compact by 2.2 and metrizable, for example by 2.10, we obtain a well defined metrizable topological space $\mathrm{Map}(S,T)$. It follows easily from the Weierstrass Theorem 1.4 that the space $\mathrm{Hol}(S,T)$ of holomorphic maps is a closed subset of $\mathrm{Map}(S,T)$.

> **Theorem 3.2. Hyperbolic Compactness.** *If S and T are hyperbolic Riemann surfaces, then the space $\mathrm{Hol}(S,T)$ of holomorphic maps is locally compact and σ-compact. Furthermore, if $K \subset S$ and $K' \subset T$ are non-vacuous compact subsets, then the set of all holomorphic maps $f : S \to T$ satisfying $f(K) \subset K'$ forms a compact subset of $\mathrm{Hol}(S,T)$.*

In particular, if T itself is compact, it follows that the entire space

$\text{Hol}(S, T)$ is compact. More generally, if $k_0 \in S$ is any base point, it follows that the evaluation map $f \mapsto f(k_0)$ is a *proper map*

$$\text{Hol}(S, T) \;\to\; T \,.$$

That is, the pre-image of any compact set $K' \subset T$ is a compact subset of $\text{Hol}(S, T)$.

Note that these statements are clearly false in the non-hyperbolic case. For example if $S = T$ is either the Riemann sphere $\widehat{\mathbb{C}}$ or the complex numbers $\mathbb{C}$ or a quotient $\mathbb{C}/\mathbb{Z}$ or $\mathbb{C}/\Lambda$, then the sequence of maps $z \mapsto nz$ in $\text{Hol}(S, S)$ takes the compact set $K = K' = \{0\}$ into itself, and yet has no convergent subsequence since the first derivatives at zero do not converge. The space $\text{Hol}(\widehat{\mathbb{C}}, \widehat{\mathbb{C}})$ of all rational maps is actually locally compact, but not compact, while the space $\text{Hol}(\mathbb{C}, \mathbb{C})$ is not even locally compact. (Problem 3-c.)

Proof of 3.2. The proof will be based on the *Bolzano-Weierstrass Theorem*, which asserts that a metric space is compact if and only if every infinite sequence of points in the space possesses a convergent subsequence. (Problem 3-d.) Thus it will suffice to show that every sequence of holomorphic maps $f_n : S \to T$ with $f_n(K) \subset K'$ contains a convergent subsequence.

It follows easily from 2.2 that the Riemann surface S possesses a countable dense subset $\{s_j\}$, where we may assume that $s_1 \in K$. Since K' is compact, the sequence of image points $f_n(s_1) \in K'$ certainly contains a convergent subsequence. Thus we can first choose an infinite subsequence $\{f_{n(p)}\}$ of the f_n so that the images $f_{n(p)}(s_1)$ converge to some $t_1 \in T$. It then follows from 2.9 and 2.11 that the images $f_{n(p)}(s_2)$ of the point s_2 lie within some compact subset of T. Hence we can choose a sub-sub-sequence $f_{n(p(q))}$ so that the $f_{n(p(q))}(s_2)$ converge to some limit t_2. Now continue inductively. By a diagonal procedure, taking the first element of the first subsequence, the second element of the second subsequence, and so on, we construct a new infinite sequence of maps $g_m = f_{n_m}$ so that $\lim_{m \to \infty} g_m(s_j) = t_j \in T$ exists for each fixed j. We claim that this new sequence $\{g_m\}$ converges, uniformly on every compact subset of S, to a holomorphic map $g : S \to T$.

Given any compact set $L \subset S$ and any $\epsilon > 0$, we can cover L by finitely many open balls of radius ϵ, centered at points s_j. In other words, we can choose finitely many points from $\{s_j\}$ so that every $z \in L$ has Poincaré distance $\text{dist}_S(z, s_j) < \epsilon$ from one of these s_j. Further, we can choose n_0 so that $\text{dist}_T(g_m(s_j), g_n(s_j)) < \epsilon$ for each of these finitely many s_j, whenever $m, n > n_0$. For any $z \in L$ it then follows using 2.11

and the triangle inequality that

$$\text{dist}_T(g_m(z), g_n(z)) < 3\epsilon$$

whenever $m, n > n_0$. Thus the $g_m(z)$ form a Cauchy sequence. It follows that the sequence of functions $\{g_m|_L\}$ converges uniformly to a limit. Since L is an arbitrary compact set in S, this proves that $\{g_m\}$ converges locally uniformly to a limit function, which must belong to $\text{Hol}(S,T)$. Therefore, by Bolzano-Weierstrass, the set of all $f \in \text{Hol}(S,T)$ with $f(K) \subset K'$ is compact.

In particular, it follows that the evaluation map $f \mapsto f(k_0)$ from $\text{Hol}(S,T)$ to T is proper. Since T is locally compact and σ-compact, this implies that $\text{Hol}(S,T)$ is also locally compact and σ-compact. $\square$

Normal families. Here is a preliminary definition: A collection $\mathcal{F}$ of holomorphic functions from a Riemann surface S to a *compact* Riemann surface T is called a *normal* family if its closure $\bar{\mathcal{F}} \subset \text{Hol}(S,T)$ is a compact set, or equivalently if every infinite sequence of functions $f_n \in \mathcal{F}$ contains a subsequence which converges locally uniformly to some limit function $g : S \to T$.

We will also need to consider the case of a non-compact target surface T. For this purpose, we need the following definition: A sequence of points t_n in the non-compact surface T *diverges to infinity* in T if for every compact set $K \subset T$ we have $t_n \neq K$ for n sufficiently large. (Here the qualification "in T" is essential. For example the sequence of points i/n diverges to infinity in the upper half-plane $\mathbb{H}$, but converges to 0 in $\mathbb{C}$.) Similarly, we will say that a sequence of maps $f_n : S \to T$ *diverges locally uniformly to infinity* in T if, for every compact $K \subset S$ and $K' \subset T$, we have $f_n(K) \cap K' = \emptyset$ for n sufficiently large. (Of course this can never happen if T itself is compact.)

Definition. A collection $\mathcal{F}$ of maps from a Riemann surface S to a (possibly non-compact) Riemann surface T will be called *normal* if every infinite sequence of maps from $\mathcal{F}$ either contains a subsequence which converges locally uniformly, or a subsequence which diverges locally uniformly to infinity in T. We can now restate 3.2 as follows.

Corollary 3.3. *If S and T are hyperbolic, then every family $\mathcal{F}$ of holomorphic maps from S to T is normal.*

Proof. Choose base points $s_0 \in S$ and $t_0 \in T$. If the set of images $\{f(s_0) ; f \in \mathcal{F}\}$ lies in some compact subset $K' \subset T$, then it follows immediately from 3.2 that $\bar{\mathcal{F}}$ is compact. Otherwise, we can choose an infinite sequence of maps $f_n \in \mathcal{F}$ so that the Poincare distance $\text{dist}_T(t_0, f_n(s_0))$

tends to infinity. Using Pick's Theorem 2.11, it then follows easily that this sequence of maps f_n diverges locally uniformly to infinity in T.   □

As an application of this result, we can compare the Poincaré metrics in a pair of Riemann surfaces $S \subset S'$. (Compare (2 : 6).) We will use the notation $N_r(p) \subset S$ for the open neighborhood of radius r consisting of all $q \in S$ with $\text{dist}_S(p, q) < r$, using the S Poincaré metric.

Theorem 3.4. Poincaré metric near the boundary. *Suppose that $S \subset S'$ are Riemann surfaces with S hyperbolic, and let $p_1, p_2, \ldots$ be a sequence of points in S which converges (in the topology of S') to a boundary point $\hat{p} \in \partial S \subset S'$. Then for any fixed r the entire neighborhood $N_r(p_j)$ converges uniformly to $\hat{p}$ as $j \to \infty$. If S has compact closure in S', then choosing some metric on S' compatible with its topology, we can make the following sharper statement. The diameter in this S'-metric of the neighborhood $N_r(p_j)$ converges uniformly to zero as p_j converges towards ∂S.*

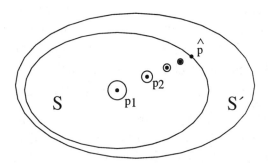

(Compare A.8 in the Appendix for a more quantitative estimate when S is a simply connected open subset of $\mathbb{C}$.)

Proof of 3.4. First note the following preliminary statement: *If K is any compact subset of S, and if $\{p_j\}$ converges to ∂S, then $N_r(p_j) \cap K = \emptyset$ for j sufficiently large.* To see this, let k_0 be a basepoint in K and let r_K be the diameter of K. Then $N_{r+r_K}(k_0)$ is compact by 2.9. For j sufficiently large, p_j will be outside of this compact set, and hence $N_r(p_j)$ will be disjoint from K.

Let $N_r^0 \subset \mathbb{D}$ be the disk of Poincaré radius r about the central point of the unit disk. If $\tilde{S}$ is the universal covering of S, then composing a suitable isomorphism $\mathbb{D} \cong \tilde{S}$ with the projection $\tilde{S} \to S$ we can construct a covering map $f_j : \mathbb{D} \to S$ with $f_j(0) = p_j$. Evidently $N_r(p_j)$ can be identified with the image $f_j(N_r^0)$ of this standard disk.

For any sufficiently large compact set $K \subset S$, note that each component of $S' \smallsetminus K$ will be a hyperbolic Riemann surface. The maps $f_j|_{N_r^0}$, for j sufficiently large, take values in $S' \smallsetminus K$, and hence form a normal family. If the p_j all lie in some compact subset of S' (for example if $\{p_j\}$ converges to some point of S'), then we can choose a subsequence so that $f_j|_{N_r^0}$ converges locally uniformly to a holomorphic limit map $f : N_r^0 \to S' \smallsetminus K$. (In fact, since we can apply the same argument to a disk of radius $r+1$, it follows that this subsequence converges uniformly on the closed disk $\overline{N}_r^0$.)

We claim that f must map the entire disk N_r^0 to a single point of $\partial S \subset S'$. For if this limit map were not constant, then its image $f(N_r^0)$ would be an open subset of $S' \smallsetminus K$. Hence this image would have to intersect S. But this is impossible since S can be exhausted by a sequence of compact subsets $K_1 \subset K_2 \subset \cdots$, and the argument above shows that $f(N_r^0)$ must be disjoint from each K_n.

The above discussion applied only to some subsequence of the $N_r(p_j)$. Now we must deal with the entire sequence. Choosing some metric $\mathrm{dist}'(p, q)$ on the space S', let d_j' be the diameter of the set $N_r(p_j) \subset S \subset S'$ with respect to this dist' metric. We must show that the sequence d_j' converges uniformly to zero. Otherwise we could choose a subsequence with $d_j' \geq \epsilon > 0$, and then choose a subsequence of this so that $f_j|_{\overline{N}_r^0}$ converges uniformly to a constant map. Evidently this is impossible. $\square$

Lemma 3.5. *Given Riemann surfaces S and $U \subset T$, let $f_j : S \to U$ be a sequence of holomorphic maps which diverges to infinity in U but not in T. Then there exists a subsequence which converges locally uniformly to a constant map from S to a single point of $\partial U \subset T$.*

Proof. Since $\{f_j\}$ does not diverge to infinity in T, we can choose compact sets $K \subset S$ and $L \subset T$ so that $f_j(K) \cap L \neq \emptyset$ for infinitely many j. After passing to a subsequence, we may choose points $k_{j_i} \in K$ so that the images $f_{j_i}(k_{j_i})$ converge to a limit $\ell \in L$.

First suppose that S and U are hyperbolic. Since K has finite diameter in the S-Poincaré metric, it follows from 3.4 and Pick's Theorem that the entire image $f_{j_i}(K)$ converges to ℓ. Again using 3.4, it follows easily that this sequence of maps $f_{j_i} : S \to U \subset T$ converges locally uniformly to the constant map $K \mapsto \ell \in \partial U \subset T$, as required.

If S and U are not hyperbolic, then we can choose a hyperbolic neighborhood S_0 of $K \subset S$ with compact closure, and a hyperbolic set U_0 of the form $U_0 = U \smallsetminus \text{compact}$. The above argument shows that a subsequence of the f_j restricted to S_0 converges locally uniformly to a constant map. Since S_0 can be arbitrarily large, this completes the proof. $\square$

Corollary 3.6. *Given Riemann surfaces S and $U \subset T$, a family of maps from S to U is normal if and only if it is normal considered as a family of maps from S into the larger surface T.*

The proof is immediate. Combining 3.6 with 3.3 and the fact that the thrice punctured sphere is hyperbolic (see 2.4), we obtain the following important consequence.

Theorem 3.7 (Montel). *Let S be a Riemann surface, and let $\mathcal{F}$ be a collection of holomorphic maps $f : S \to \widehat{\mathbb{C}}$ which omit three different values. That is, assume that there are distinct points $a, b, c \in \widehat{\mathbb{C}}$ so that $f(S) \subset \widehat{\mathbb{C}} \setminus \{a, b, c\}$ for every $f \in \mathcal{F}$. Then $\mathcal{F}$ is a normal family; that is, the closure $\overline{\mathcal{F}} \subset \mathrm{Hol}(S, \widehat{\mathbb{C}})$ is a compact set.*

Concluding Problems.

Problem 3-a. The compact-open topology. If X and Y are locally compact spaces, the *compact-open topology* on the space $\mathrm{Map}(X, Y)$ of all maps is defined to be the *smallest* topology (that is the topology with fewest open sets), such that, for every compact $K \subset X$ and every open $U \subset Y$, the set of $f : X \to Y$ with $f(K) \subset U$ forms an open subset of $\mathrm{Map}(X, Y)$. If Y is metrizable, show that this coincides with the topology of locally uniform convergence, as described above. Show that the composition operation

$$f, g \mapsto g \circ f$$

is continuous as a mapping from $\mathrm{Map}(X, Y) \times \mathrm{Map}(Y, Z)$ to $\mathrm{Map}(X, Z)$. If U is an open subset of Y, show that $\mathrm{Map}(X, U)$ embeds homeomorphically as a subset of $\mathrm{Map}(X, Y)$, but that this subset need not be either open or closed.

Now suppose that S and T are Riemann surfaces and that U is a connected open subset of T. Show that the topological boundary of $\mathrm{Hol}(S, U)$ in $\mathrm{Hol}(S, T)$ consists completely of constant maps from S to ∂U.

Problem 3-b. Uniform convergence or divergence? Consider the family of maps $f_n(z) = z + n$ from $\mathbb{C}$ or $\widehat{\mathbb{C}}$ to itself. Show that this sequence diverges locally uniformly to infinity in $\mathbb{C}$. However, in $\widehat{\mathbb{C}}$ show that this sequence neither converges nor diverges locally uniformly, although it does converge pointwise to the constant function which maps all of $\widehat{\mathbb{C}}$ to the single point $\infty \in \widehat{\mathbb{C}}$. Similarly, show that the sequence of degree one rational functions $g_n(z) = 1/(n^2 z - n)$ converges pointwise, but not locally uniformly, to the constant function $g(z) = 0$.

Problem 3-c. Local compactness? Show that $\mathrm{Hol}(\mathbb{C}, \mathbb{C})$ is not locally compact, since every neighborhood of the zero map contains a sequence of polynomial maps of the form

$$f_n(z) \;=\; \epsilon(1 + \epsilon z + \epsilon^2 z^2 + \cdots + \epsilon^n z^n)$$

with no limit point in $\mathrm{Hol}(\mathbb{C}, \mathbb{C})$. Similarly show that $\mathrm{Hol}(\mathbb{C}, \widehat{\mathbb{C}})$ is not locally compact. However, if S and T are compact, show that $\mathrm{Hol}(S, T)$ is always locally compact.[*]

Problem 3-d. Bolzano-Weierstrass. If X is a metric space with the property that every infinite sequence possesses an accumulation point (or equivalently a convergent subsequence), show that, for each $\epsilon > 0$, X can be covered by finitely many balls of radius ϵ. If X is the union of open subsets U_α, conclude that X can actually be expressed as the union of finitely many of the U_α; in other words, X is compact.

Problem 3-e. Local normality. Show that normality is a local property. More precisely, let S and T be any Riemann surfaces, and let $\{f_\alpha\}$ be a family of holomorphic maps from S to T. If every point of S has a neighborhood U such that the collection $\{f_\alpha|_U\}$ of restricted maps is a normal family in $\mathrm{Hol}(U, T)$, show by a diagonal argument, similar to the proof of 3.2, that the family $\{f_\alpha\}$ itself is normal.

Problem 3-f. Normality and derivatives. Let $f : S \to T$ be holomorphic. Given Riemannian metrics on the Riemann surfaces S and T, we can define the *norm* of the derivative of f at a point $s \in S$ to be the real number $\|f'(s)\| \geq 0$ such that the induced linear mapping from the tangent space of S at s to the tangent space of T at $f(s)$ carries vectors of length 1 to vectors of length $\|f'(s)\|$. If T is compact, show that a family $\mathcal{F}$ of maps $f : S \to T$ is normal if and only if the collection

---

[*] The space $\mathrm{Hol}(S, \widehat{\mathbb{C}})$ of meromorphic functions on S is of particular interest. As an example, for each $d \geq 1$ the space $\mathrm{Hol}_d(\widehat{\mathbb{C}}, \widehat{\mathbb{C}})$ of degree d rational maps forms a complex $(2d + 1)$-dimensional manifold which is non-compact. (See for example Segal (1979) or Milnor (1993).) On the other hand, there exists a surface S of genus five so that the space $\mathrm{Hol}(S, \widehat{\mathbb{C}})$ has singularities. (Private communication from J. Harris.) If the target space T is a torus, then each connected component of $\mathrm{Hol}(S, T)$ is a copy of T, while if T has genus ≥ 2 then $\mathrm{Hol}(S, T)$ is a finite set.

of norms $\|f'(s)\|$ is uniformly bounded as f varies over $\mathcal{F}$ and s varies over any compact subset of S.

Problem 3-g. The one point compactification. For any locally compact space X, let $X \cup \infty$ be the topological space which is obtained by adjoining a single point ∞ to X, defining the basic neighborhoods of ∞ to be the complements of compact subsets of X. Show that $X \cup \infty$ is a compact Hausdorff space, metrizable if X is metrizable and σ-compact.

Now let S and T be non-compact Riemann surfaces. Show that the closure of $\mathrm{Hol}(S,T)$ in the larger space $\mathrm{Map}(S, T \cup \infty)$ consists of $\mathrm{Hol}(S,T)$ together with the constant map $[\infty]$ which carries all of S to the point ∞. If S and T are hyperbolic, show that this closure

$$\mathrm{Hol}(S,T) \cup [\infty] \;\subset\; \mathrm{Map}(S, T \cup \infty)$$

is compact, and can be identified with the one point compactification of $\mathrm{Hol}(S,T)$.

ITERATED HOLOMORPHIC MAPS

§4. Fatou and Julia: Dynamics on the Riemann Sphere

The local study of iterated holomorphic mappings, in a neighborhood of a fixed point, was quite well developed in the late 19th century. (Compare §§8-10, and see Alexander.) However, except for one very simple case studied by Schröder and Cayley (see Problem 7-a), nothing was known about the global behavior of iterated holomorphic maps until 1906, when Pierre Fatou described the following startling example. For the map $z \mapsto z^2/(z^2 + 2)$, he showed that almost every orbit under iteration converges to zero, even though there is a Cantor set of exceptional points for which the orbit remains bounded away from zero. (Problems 4-e, f.) This aroused great interest. After a hiatus during the first world war, the subject was taken up in depth by Fatou, and also by Gaston Julia and others such as S. Lattès and J. F. Ritt. The most fundamental and incisive contributions were those of Fatou himself. However Julia was a determined competitor, and tended to get more credit because of his status as a wounded war hero. (In 1918, Julia was awarded the "Grand Prix des Sciences Mathématiques" by the Paris Academy of Sciences for his work.)

Definition. Let S be a compact Riemann surface, let $f : S \to S$ be a non-constant holomorphic mapping, and let $f^{on} : S \to S$ be its n-fold iterate. The domain of normality for the collection of iterates $\{f^{on}\}$ is called the *Fatou set*[*] Fatou(f), and its complement $S \smallsetminus$ Fatou(f) is called the *Julia set* $J = $ Julia(f). Thus, for any point $p_0 \in S$, we have the following basic dichotomy: *If there exists some neighborhood U of p_0 so that the sequence of iterates $\{f^{on}\}$ restricted to U forms a normal family of maps from U to S, then we say that p_0 belongs to the Fatou set of f. Otherwise, if no such neighborhood exists, we say that p_0 belongs to the Julia set.*

Thus, by its very definition, the Julia set J is a closed subset of S, while the Fatou set $S \smallsetminus J$ is the complementary open subset. We will see that a point p_0 belongs to the Julia set if and only if dynamics in a neighborhood of p_0 displays "*sensitive dependence on initial conditions*", so that nearby initial conditions lead to wildly different behavior after a large

---

[*]   The choice as to which of these two sets should be named after Julia and which after Fatou is rather arbitrary, but the term "Julia set" now seems firmly established. (Note however that this form of the definition of J is actually due to Fatou. For Julia's definition, see §14.) The Fatou set $S \smallsetminus J$ is sometimes called by other names such as "stable set" or "normal set".

(or sometimes not so large) number of iterations. (Compare Problem 4-h, as well as 14.2.)

The classical example, and the one which we will emphasize, is the case where S is the Riemann sphere $\widehat{\mathbb{C}} = \mathbb{C} \cup \infty$. Any holomorphic map $f : \widehat{\mathbb{C}} \to \widehat{\mathbb{C}}$ on the Riemann sphere can be expressed as a rational function, that is as the quotient $f(z) = p(z)/q(z)$ of two polynomials. Here we may assume that $p(z)$ and $q(z)$ have no common roots. The *degree* d of $f = p/q$ is then equal to the maximum of the degrees of p and q. For all but finitely many choices of constant $c \in \widehat{\mathbb{C}}$, this degree can be described as the number of distinct solutions to the equation $f(z) = c$. We will usually assume that $d \geq 2$, and always that $d \geq 1$ so that f is a non-constant map from $\widehat{\mathbb{C}}$ onto itself.

As a simple example, consider the squaring map $s : z \mapsto z^2$ on $\widehat{\mathbb{C}}$. The entire open disk $\mathbb{D}$ is contained in the Fatou set of s, since successive iterates on any compact subset converge uniformly to zero. Similarly the exterior $\widehat{\mathbb{C}} \smallsetminus \overline{\mathbb{D}}$ is contained in the Fatou set, since the iterates of s converge to the constant function $z \mapsto \infty$ outside of $\overline{\mathbb{D}}$. On the other hand, if z_0 belongs to the unit circle, then in any neighborhood of z_0 any limit of iterates $s^{\circ n}$ would necessarily have a jump discontinuity as we cross the unit circle. This shows that the Julia set $J(s)$ is precisely equal to the unit circle.

Such smooth Julia sets are rather exceptional. (See §7.) Figure 1 shows some more typical examples of Julia sets (colored black) for quadratic polynomial mappings. Figure 1a shows a rather wild Jordan curve, Figure 1b a rather thick Cantor set, Figure 1c a "dendrite", and Figure 1d a Julia set which cuts the plane into infinitely many "Fatou components". (The arrows give a rough indication of what maps to what.) In each of these pictures, since $f(z)$ is an even function, the Julia set is centrally symmetric. Three examples of non-polynomial Julia sets are shown in Figure 2. (Compare McMullen (1988) for the first, Milnor (1993) for the second, and Bleher and Lyubich for the third.)

Here are some basic properties of the Julia set.

4.1. Invariance Lemma. *The Julia set $J = $ Julia(f) of a holomorphic map $f : S \to S$ is **fully invariant** under f. That is z belongs to J if and only if the image $f(z)$ belongs to J.*

A completely equivalent statement is that the Fatou set is fully invariant. In fact, for any open set $U \subset S$, some sequence of iterates $f^{\circ n_j}$ converges uniformly on compact subsets of U if and only if the corresponding sequence of iterates $f^{\circ n_j + 1}$ converges uniformly on compact subsets of

Figure 1a. A simple closed curve,
Julia set for $z \mapsto z^2 + (.99 + .14i)z$.

Figure 1b. A totally disconnected Julia set,
$z \mapsto z^2 + (-.765 + .12i)$.

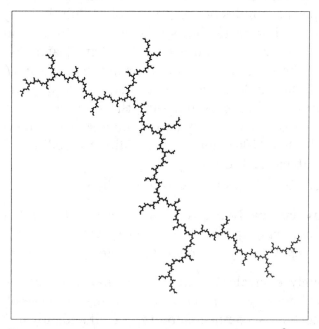

Figure 1c. A "dendrite", Julia set for $z \mapsto z^2 + i$.

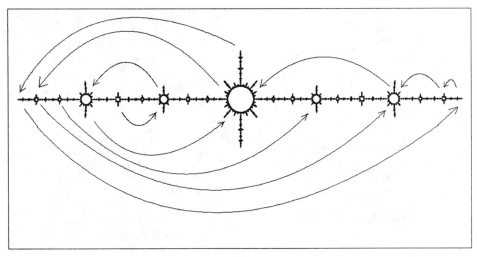

Figure 1d. Julia set for $z \mapsto z^2 - 1.75488\ldots$, the "airplane".

the open set $f^{-1}(U)$. Further details will be left to the reader. $\square$

It follows that the Julia set possesses a great deal of *self-similarity*: *Whenever* $f(z_1) = z_2$ *in* $J(f)$, *with derivative* $f'(z_1) \neq 0$, *there is an induced conformal isomorphism from a neighborhood* N_1 *of* z_1 *to a neighborhood* N_2 *of* z_2 *which takes* $N_1 \cap J(f)$ *precisely onto* $N_2 \cap J(f)$. (Compare Problem 4-d.)

4.2. Iteration Lemma. *For any* $k > 0$, *the Julia set* $J(f^{\circ k})$ *of the* k*-fold iterate coincides with the Julia set* $J(f)$.

Proof outline. Again we can equally well work with the set $\mathrm{Fatou}(f) = S \smallsetminus J$. Suppose, for example, that z belongs to the Fatou set of $f \circ f$. This means that, for some neighborhood U of z, the collection of all even iterates $f^{\circ 2n}|_U$ is contained in a compact subset $K \subset \mathrm{Hol}(U, S)$. It follows that every iterate of f, restricted to U, belongs to the compact set $K \cup (f \circ K) \subset \mathrm{Hol}(U, S)$, hence z belongs to the Fatou set of f. Further details will be left to the reader. $\square$

Definitions. Consider a *periodic orbit* or "*cycle*"

$$f : z_0 \mapsto z_1 \mapsto \cdots \mapsto z_{m-1} \mapsto z_m = z_0$$

for a holomorphic map $f : S \to S$. (Here S can be any Riemann surface, compact or not.) If the points $z_1, \ldots, z_m$ are all distinct, then the integer $m \geq 1$ is called the *period*. The first derivative of the m-fold iterate $f^{\circ m}$ at a point of the orbit is a well defined complex number called the *multiplier* of the orbit. If the Riemann surface S is an open subset of $\mathbb{C}$,

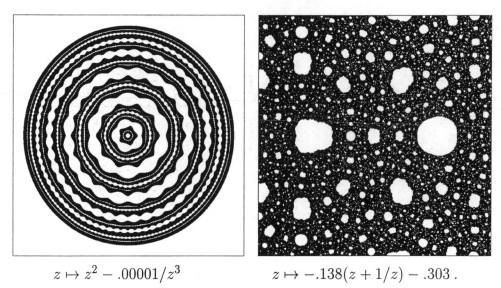

$$z \mapsto z^2 - .00001/z^3 \qquad\qquad z \mapsto -.138(z + 1/z) - .303 \ .$$

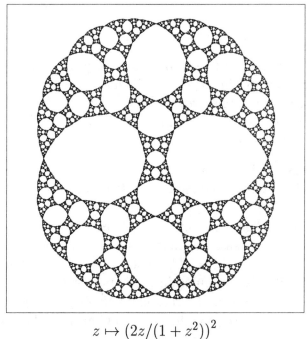

$$z \mapsto (2z/(1 + z^2))^2$$

Fig 2. Julia sets for three rational maps.

then we have the product formula

$$\lambda \; = \; (f^{\circ m})'(z_i) \; = \; f'(z_1) \cdot f'(z_2) \cdots f'(z_m) \ .$$

In particular, $\lambda = 0$ if and only if some point z_j of the orbit is a *critical point* of f, that is, a point at which the first derivative f' vanishes. More generally, for self-maps of an arbitrary Riemann surface we have a corresponding product formula, using a local uniformizing parameter (= local coordinate chart) around each point of the orbit. The product λ is independent of the choice of uniformizing parameters. By definition, a periodic orbit is either *attracting* or *repelling* or *indifferent* (= *neutral*) according as its multiplier satisfies $|\lambda| < 1$ or $|\lambda| > 1$ or $|\lambda| = 1$. (Compare §8.) The orbit will be called *superattracting* if $\lambda = 0$, and *geometrically attracting* if $0 < |\lambda| < 1$. As an example, the map illustrated in Figure 1d has a superattracting orbit of period three.

Caution: In the special case where the point at infinity is periodic under a rational map, $f^{\circ m}(\infty) = \infty$, this definition may be confusing. The multiplier λ is *not* equal to the limit as $z \to \infty$ of the derivative of $f^{\circ m}(z)$, but rather turns out to be equal to the reciprocal of this number. (Problem 4-c.) As examples, if $f(z) = 2z$ then ∞ is an attracting fixed point with multiplier $\lambda = 1/2$, while if f is a polynomial of degree $d \geq 2$ then ∞ is a superattracting fixed point, with $\lambda = 0$.

Definition. If $\mathcal{O}$ is an attracting periodic orbit of period m, we define the *basin of attraction* to be the open set $\mathcal{A} \subset S$ consisting of all points $z \in S$ for which the successive iterates $f^{\circ m}(z)$, $f^{\circ 2m}(z)$, ... converge towards some point of $\mathcal{O}$. Assuming once more that S is compact, we have the following.

4.3. Lemma. Basins and repelling points. *Every attracting periodic orbit is contained in the Fatou set of f. In fact the entire basin of attraction $\mathcal{A}$ for an attracting periodic orbit is contained in the Fatou set. However every repelling periodic orbit is contained in the Julia set.*

Proof of 4.3. First consider a fixed point $z_0 = f(z_0)$ with multiplier λ. If $|\lambda| > 1$, then no sequence of iterates of f can converge uniformly near z_0. For the first derivative of $f^{\circ n}$ at z_0 is λ^n, which diverges to infinity as $n \to \infty$. (Compare the Weierstrass Uniform Convergence Theorem 1.4.) On the other hand, if $|\lambda| < 1$, then choosing $|\lambda| < c < 1$ it follows from Taylor's Theorem that $|f(z) - z_0| \leq c|z - z_0|$ for z sufficiently close to z_0, hence the successive iterates of f, restricted to a small neighborhood, converge uniformly to the constant function $z \mapsto z_0$. (See 8.1 for details.) The corresponding statement for any compact subset of the basin $\mathcal{A}$ then follows easily. These statements for fixed points generalize immediately to periodic points, making use of 4.2, since a periodic point of f is just a fixed point of some iterate $f^{\circ m}$. □

The case of an indifferent periodic point is much more difficult. (Compare §§10, 11.) One particularly important case is the following.

Definition. A periodic point $z_0 = f^{\circ n}(z_0)$ is called *parabolic* if the multiplier λ at z_0 is equal to $+1$, yet $f^{\circ n}$ is not the identity map, or more generally if λ is a root of unity, yet no iterate of f is the identity map.

As an example, the two fixed points of the rational map $f(z) = z/(z - 1)$ both have multiplier equal to -1. However, these do not count as parabolic points since $f \circ f(z)$ is identically equal to z. We must exclude such cases so that the following will be true.

> **4.4. Lemma. Parabolic points.** *Every parabolic periodic point belongs to the Julia set.*

Proof. Let w be a local uniformizing parameter, with $w = 0$ corresponding to the periodic point. Then some iterate $f^{\circ m}$ corresponds to a local mapping of the w-plane with power series expansion of the form $w \mapsto w + a_q w^q + a_{q+1} w^{q+1} + \cdots$, where $q \geq 2$, $a_q \neq 0$. It follows that $f^{\circ mk}$ corresponds to a power series $w \mapsto w + k a_q w^q + \cdots$. Thus the q-th derivative of $f^{\circ mk}$ at 0 is equal to $q! \, k \, a_q$, which diverges to infinity as $k \to \infty$. It follows from 1.4 that no subsequence $\{f^{\circ mk_j}\}$ can converge locally uniformly as $k_j \to \infty$. $\square$

Now and for the rest of §4, let us specialize to the case of a rational map $f : \widehat{\mathbb{C}} \to \widehat{\mathbb{C}}$ of degree $d \geq 2$.

> **4.5. Lemma. J is not empty.** *If f is rational of degree two or more, then the Julia set $J(f)$ is non-vacuous.*

Proof. If $J(f)$ were vacuous, then some sequence of iterates $f^{\circ n_j}$ would converge, uniformly over the entire sphere $\widehat{\mathbb{C}}$, to a holomorphic limit $g : \widehat{\mathbb{C}} \to \widehat{\mathbb{C}}$. Here we are using the fact that normality is a local property (Problem 3-e). A standard topological argument would then show that the degree of the map $f^{\circ n_j}$ is equal to the degree of g for large j. (In fact if two maps f_j and g are sufficiently close that the spherical distance $\sigma(f_j(z), g(z))$ is uniformly less than the distance π between antipodal points, then we can deform $f_j(z)$ to $g(z)$ along the unique shortest geodesic; hence these two maps are homotopic and have the same degree.) But the degree of $f^{\circ n}$ cannot equal the degree of g for large n, since the degree of $f^{\circ n}$ is equal to d^n, which diverges to infinity as $n \to \infty$. $\square$

A different, more constructive proof of this Lemma will be given in 12.5.

We will also need the following concepts.

Definition. By the *grand orbit* of a point z under $f : S \to S$ we mean the set $\mathbf{go}(z, f)$ consisting of all points $z' \in S$ whose orbits eventually intersect the orbit of z. Thus z and z' have the same grand orbit if and only if $f^{\circ m}(z) = f^{\circ n}(z')$ for some choice of $m \geq 0$ and $n \geq 0$. A point $z \in S$ will be called **grand orbit finite** or (to use the classical terminology) *exceptional* under f if its grand orbit $\mathbf{go}(z, f) \subset S$ is a finite set. Using Montel's Theorem, we prove the following.

4.6. Lemma. Finite grand orbits. *If $f : \widehat{\mathbb{C}} \to \widehat{\mathbb{C}}$ is rational of degree $d \geq 2$, then the set $\mathcal{E}(f)$ of grand orbit finite points can have at most two elements. These grand orbit finite points, if they exist, must always be superattracting periodic points of f, and hence must belong to the Fatou set.*

Proof. (Compare Problem 4-b.) Since f maps $\widehat{\mathbb{C}}$ onto itself, it must map any grand orbit $\mathbf{go}(z, f)$ onto itself. Hence, if this grand orbit is finite, it must map bijectively onto itself, and hence constitute a single periodic orbit $a_0 \mapsto a_1 \mapsto \cdots \mapsto a_m = a_0$. Now note that an arbitrary point $\hat{z} \in \widehat{\mathbb{C}}$ has exactly d preimages under f, counted with multiplicity, where the *multiplicity* of $z_j \in f^{-1}(z)$ as a preimage is greater than one if and only if z_j is a critical point of f. Setting $f(z) = p(z)/q(z)$ (and assuming that ∞, $f(\infty) \neq \hat{z}$), this is just a matter of counting roots of the degree d polynomial equation $p(z) - \hat{z}q(z) = 0$, checking that the derivative of f vanishes at any multiple root. It follows that every a_j in this finite periodic orbit must be a critical point of f. This proves that any finite grand orbit is superattracting, and hence contained in the Fatou set.

(Caution: This argument makes strong use of the fact that every non-constant map from $\widehat{\mathbb{C}}$ to itself is onto. An entire map from $\mathbb{C}$ to itself, for example $z \mapsto 2ze^z$, may well have a repelling point which is grand orbit finite. See Problem 6-c.)

If there were three distinct grand orbit finite points, then the union of the grand orbits of these points would form a finite set whose complement U in $\widehat{\mathbb{C}}$ would be hyperbolic, with $f(U) = U$. Therefore, the set of iterates of f restricted to U would be normal by Montel's Theorem. Hence both U and its complement would be contained in the Fatou set, contradicting 4.5. $\square$

4.7. Transitivity Theorem. *Let z_1 be an arbitrary point of the Julia set $J(f) \subset \widehat{\mathbb{C}}$ and let N be an arbitrary neighborhood of z_1. Then the union U of the forward images $f^{\circ n}(N)$ contains the entire Julia set, and contains all but at most two points of $\widehat{\mathbb{C}}$. More precisely, if N is sufficiently small, then U is the complement $\widehat{\mathbb{C}} \smallsetminus \mathcal{E}(f)$ of the set of grand orbit finite points.*

(In §14 we will prove the much sharper statement that a single forward image $f^{\circ n}(N)$ actually contains the entire Julia set, or the entire Riemann sphere in the special case where there are no grand orbit finite points, provided that n is sufficiently large.)

Proof of 4.7. First note that the complementary set $\widehat{\mathbb{C}} \smallsetminus U$ can contain at most two points. For otherwise, since $f(U) \subset U$, it would follow from Montel's Theorem that U must be contained in the Fatou set, which is impossible since $z_1 \in U \cap J$. Again making use of the fact that $f(U) \subset U$, we see that any preimage of a point $z \in \widehat{\mathbb{C}} \smallsetminus U$ must itself belong to the finite set $\widehat{\mathbb{C}} \smallsetminus U$. It follows by a counting argument that some iterated preimage of z is periodic; hence z itself is periodic and grand orbit finite. Since the set $\mathcal{E}(f)$ of grand orbit finite points is disjoint from J, it follows that $J \subset U$. Finally, if N is small enough so that $N \subset \widehat{\mathbb{C}} \smallsetminus \mathcal{E}(f)$, it follows easily that $U = \widehat{\mathbb{C}} \smallsetminus \mathcal{E}(f)$. $\square$

4.8. Corollary. Julia set with interior. *If the Julia set contains an interior point, then it must be equal to the entire Riemann sphere.*

For if $J = J(f)$ has an interior point z_1 then, choosing a neighborhood $N \subset J$ of z_1, the union $U \subset J$ of forward images of N is everywhere dense, $\overline{U} = \widehat{\mathbb{C}}$. Since J is a closed set, it follows that $J = \widehat{\mathbb{C}}$. (For examples, see §7.) $\square$

4.9. Corollary. Basin boundary = Julia set. *If $\mathcal{A} \subset \widehat{\mathbb{C}}$ is the basin of attraction for some attracting periodic orbit, then the topological boundary $\partial \mathcal{A} = \overline{\mathcal{A}} \smallsetminus \mathcal{A}$ is equal to the entire Julia set. Every connected component of the Fatou set $\widehat{\mathbb{C}} \smallsetminus J$ either coincides with some connected component of this basin $\mathcal{A}$ or else is disjoint from $\mathcal{A}$.*

Proof. If N is any neighborhood of a point of the Julia set, then 4.7 implies that some $f^{\circ n}(N)$ intersects $\mathcal{A}$, hence N itself intersects $\mathcal{A}$. This proves that $J \subset \overline{\mathcal{A}}$. But J is disjoint from $\mathcal{A}$, so it follows that $J \subset \partial \mathcal{A}$. On the other hand, if N is a neighborhood of a point of $\partial \mathcal{A}$, then any limit of iterates $f^{\circ n}|_N$ must have a jump discontinuity between $\mathcal{A}$ and $\partial \mathcal{A}$, hence $\partial \mathcal{A} \subset J$. Finally, note that any connected Fatou component which intersects $\mathcal{A}$, since it cannot intersect the boundary of $\mathcal{A}$, must coincide with some component of $\mathcal{A}$. $\square$

Caution. $\partial \mathcal{A}$ is not the same thing as the union of the boundaries of the connected components of $\mathcal{A}$, which is often much smaller. (Compare Figures 1d and 2. It may be instructive to think of a Cantor set in the line, which is uncountably infinite although the union of the boundaries of its complementary intervals is countable.)

4.10. Corollary. Iterated preimages are dense. *If z_0 is any point of the Julia set $J(f)$, then the set of all iterated preimages*

$$\{z \in \widehat{\mathbb{C}} \; ; \; f^{\circ n}(z) = z_0 \quad \text{for some} \quad n \geq 0\}$$

is everywhere dense in $J(f)$.

For $z_0 \notin \mathcal{E}(f)$, so 4.7 shows that every point $z_1 \in J(f)$ can be approximated arbitrarily closely by points z whose forward orbits contain z_0. $\square$

Remark on computer graphics. (Compare Appendix G.) This Corollary suggests an algorithm for computing pictures of the Julia set: Starting with any $z_0 \in J(f)$, first compute all z_1 with $f(z_1) = z_0$, then for each such z_1 compute all z_2 with $f(z_2) = z_1$, and so on, thus eventually coming arbitrarily close to every point of $J(f)$. This method is most often used in the quadratic case, since quadratic equations are very easy to solve, and since the number d^n of n-fold iterated preimages is smallest when the degree d is two. The method is very insensitive to round-off errors, since f tends to be expanding on its Julia set, so that f^{-1} tends to be contracting. (Compare Problems 4-e, 4-f, as well as §19.) However, it does have disadvantages: the number d^n grows very rapidly with n, yet it may take a great many iterated preimages to get close to certain points of J.

4.11. Corollary. No isolated points. *If f has degree two or more, then $J(f)$ has no isolated points.*

Proof. First note that $J(f)$ must be an infinite set. For if $J(f)$ were finite it would consist of grand orbit finite points, contradicting 4.6. Hence $J(f)$ contains at least one limit point z_0. Now the iterated pre-images of z_0 form a dense set of non-isolated points in $J(f)$. $\square$

4.12. Corollary. Julia components. *For any rational map of degree two or more, the Julia set is either connected or else has uncountably many connected components.*

Proof. Suppose that $J = J(f)$ can be expressed as the union $J_0 \cup J_1$ of two disjoint, non-vacuous compact subsets. Note that both of these subsets must be infinite, since J has no isolated points. We will first show that neither of these subsets can be connected. Choose an open set U which intersects J_0 but not J_1. We claim that some forward image $f^{\circ n}(U)$ must intersect both J_0 and J_1. Otherwise, given any infinite

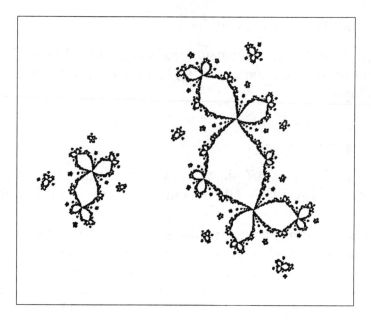

Figure 3. A family of rabbits: Julia set for a cubic polynomial,

$$f(z) \;=\; z^3 - .48z + (.706260 + .502896i) \;,$$

with infinitely many non-trivial connected components.

sequence of iterates $f^{\circ n}$, we could choose an infinite subsequence of iterates $f^{\circ n_j}$ which map U to a set which misses one of the J_α, hence omits three distinct points of $\hat{\mathbb{C}}$, and therefore has a subsequence which converges locally uniformly throughout U. This would contradict the hypothesis that U intersects the Julia set. Now choosing n so that $f^{\circ n}(U)$ intersects both J_0 and J_1, it follows from 4.1 that $f^{\circ n}(J_0)$ also intersects both J_0 and J_1. Therefore J_0 can be expressed as the disjoint union of non-empty compact subsets $J_{00} = J_0 \cap f^{-n}(J_0)$ and $J_{01} = J_0 \cap f^{-n}(J_1)$.

Similarly, it follows by induction on k that for any sequence $\alpha_1 \cdots \alpha_k$ of bits we can construct a compact non-vacuous set $J_{\alpha_1 \cdots \alpha_k}$ so that $J_{\alpha_1 \cdots \alpha_{k-1}}$ is the union of disjoint subsets $J_{\alpha_1 \cdots \alpha_k 0}$ and $J_{\alpha_1 \cdots \alpha_k 1}$. The corresponding infinite intersections

$$J_{\alpha_1 \alpha_2 \alpha_3 \cdots} \;=\; \bigcap_k J_{\alpha_1 \cdots \alpha_k}$$

are disjoint and non-vacuous, and each one contains at least one connected component of J. $\square$

Remark. In the polynomial case, it seems possible that all but countably many of these connected components must be single points. (Compare Branner and Hubbard, 1992.) However, this is certainly not true for arbitrary rational maps. (See McMullen's example, Figure 2a.)

For the last corollary, we will need some definitions. A topological space X is called a *Baire space* if every countable intersection of dense subsets of X is again dense. We will make use of *Baire's Theorem*, which asserts that every complete metric space is a Baire space, and also that every locally compact space is a Baire space. (Compare Problem 4-j.) It will be convenient to say that a property of points in the Baire space X is true for *generic* $x \in X$ if it is true for all points in some countable intersection of dense open subsets of X. We apply this concept to the topological space $J(f)$.

4.13. Corollary. Topological transitivity. *For a generic choice of the point $z \in J = J(f)$, the forward orbit*

$$\{z, f(z), f^{\circ 2}(z), \dots\}$$

is everywhere dense in J.

Proof. (Compare Problem 4-j.) For each integer $i > 0$, we can cover the Julia set $J = J(f)$ by finitely many open sets N_{ij} of diameter less than $1/i$, using the spherical metric. For each such N_{ij}, let U_{ij} be the union of the iterated pre-images $f^{-n}(N_{ij})$. It follows from 4.10 that the closure $\overline{U_{ij} \cap J}$ is equal to the entire Julia set J. In other words, $U_{ij} \cap J$ is a dense open subset of the Julia set. Now if z belongs to the intersection of these dense open sets, then the forward orbit of z intersects every one of the N_{ij}, and hence is everywhere dense in J. $\square$

Concluding Problems.

Problem 4-a. Degree one. If $f : \widehat{\mathbb{C}} \to \widehat{\mathbb{C}}$ is rational of degree $d = 1$, show that the Julia set $J(f)$ is either vacuous, or consists of a single repelling or parabolic fixed point.

Problem 4-b. Maps with grand orbit finite points. Now suppose that f is rational of degree $d \geq 2$. Show that f is actually a polynomial if and only if $f^{-1}(\infty) = \{\infty\}$, so that the point at infinity is a grand orbit finite fixed point for f. Show that f has both zero and infinity as grand orbit finite points if and only if $f(z) = \alpha z^n$, where $n = \pm d$ and $\alpha \neq 0$. Conclude that f has grand orbit finite points if and only if it is conjugate, under some fractional linear change of coordinates, either to a polynomial or to the map $z \mapsto 1/z^d$.

Problem 4-c. Fixed point at infinity. If f is a rational function with a fixed point at infinity, show that the multiplier λ at infinity is equal to $\lim_{z \to \infty} 1/f'(z)$. In particular, this fixed point is superattracting if and only if $f'(z) \to \infty$ as $z \to \infty$. (Take $\zeta = 1/z$ and use the series expansion $1/f(1/\zeta) = \lambda\zeta + a_2\zeta^2 + a_3\zeta^3 + \cdots$ in some neighborhood of $\zeta = 0$.)

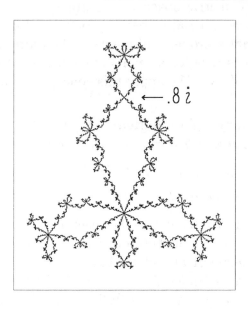

Figure 4. Julia set for $f(z) = z^3 + \frac{12}{25}z + \frac{116}{125}i$.

Problem 4-d. Self-similarity. With rare exceptions, any shape which is observed about one point of the Julia set will be observed infinitely often, throughout the Julia set. More precisely, for two points z and z' of $J = J(f)$, let us say that (J, z) is *locally conformally isomorphic* to (J, z') if there exists a conformal isomorphism from a neighborhood N of z onto a neighborhood N' of z' which carries z to z' and $J \cap N$ onto $J \cap N'$. Using 4.10, show that the set of z for which (J, z) is locally conformally isomorphic to (J, z_0) is everywhere dense in J unless the following very exceptional condition is satisfied: *For every backwards orbit*

$$z_0 \;\hookleftarrow\; z_1 \;\hookleftarrow\; z_2 \;\hookleftarrow\; \cdots$$

under f which terminates at z_0, some z_j with $j > 0$ must be a critical point of f. As an example, for the map $f(z) = z^2 - 2$ studied in §7, show that this condition is satisfied for the endpoints $z_0 = \pm 2$. Similarly, show that it is satisfied for the point $z_0 = .8i$ of Figure 4. For any f, show that there can be only finitely many such exceptional points z_0.

Problem 4-e. A Cantor Julia set. By definition, a topological space is called a *Cantor set* if it is homeomorphic to the standard middle third Cantor set $K \subset [0,1]$ consisting of all infinite sums $\sum_1^\infty 2a_i/3^i$ with coefficients $a_i \in \{0,1\}$. (If X is a compact metric space, then a standard theorem asserts that X is a Cantor set if and only if it is totally disconnected (no connected subsets other than points), with no isolated points. Compare [Hocking and Young].)

If $f(z) = z^2 - 6$, show that $J(f)$ is a Cantor set contained in the intervals $[-3, -\sqrt{3}] \cup [\sqrt{3}, 3]$. More precisely, show that a point in $J(f)$ with orbit $z_0 \mapsto z_1 \mapsto \cdots$ is uniquely determined by the sequence of signs $\epsilon_j = z_j/|z_j| = \pm 1$. In fact

$$z_0 = \epsilon_0\sqrt{6 + \epsilon_1\sqrt{6 + \epsilon_2\sqrt{6 + \cdots}}}\ .$$

(Use 4.10 and the fact that the branch $z \mapsto \sqrt{6+z}$ of f^{-1} is a strictly contracting map which carries the interval $[-3, 3]$ onto $[\sqrt{3}, 3]$. Compare Problem 2-j.) Using 4.3, show that every orbit outside of this Cantor set must escape to infinity.

Problem 4-f. The Fatou example. Similarly, for $f(z) = z^2 + c$ where $c > 1/4$ is real, show that J is a Cantor set disjoint from the real axis, and that each orbit $z_0 \mapsto z_1 \mapsto \cdots$ is uniquely determined by the sequence of signs $\epsilon_n = \mathrm{sgn}(\mathrm{Im}(z_n))$. In fact

$$z_0 = \lim_{n \to \infty} \epsilon_0 g(\epsilon_1 g(\epsilon_2 \cdots \epsilon_{n-1}g(\epsilon_n \hat{z}) \cdots))\ ,$$

where $g(z) = \sqrt{z - c}$ is the branch of f^{-1} which maps the slit plane $U = \mathbb{C} \setminus [c, +\infty)$ onto the upper half-plane, and $\hat{z}$ is some fixed base point in J, for example the fixed point in the upper half-plane. (Use the fact that g restricted to the compact set $J \subset U$ is a strictly contracting map for the Poincaré metric ρ_U.) Show that every orbit outside of J must escape to infinity. Using the substitution $w = c/z$, prove corresponding statements for Fatou's 1906 example

$$w \mapsto \frac{w^2}{c + w^2}\ .$$

Problem 4-g. Newton's method.* (See also Problem 7-a.) Let U be an open subset of $\mathbb{C}$, and let $F : U \to \mathbb{C}$ be a holomorphic map with

---

* Ever since the work of Cayley and Schröder in the 19-th century, the problem of understanding Newton's method has been a primary inspiration for the study of iterated rational functions. For recent work, see for example Tan Lei (1997), Roesch (1998), and compare Keen (1989).

derivative F'. Newton's method of searching for solutions to the equation $F(\hat{z}) = 0$ can be described as follows. Consider the auxiliary function $N : U \to \hat{\mathbb{C}}$, where

$$N(z) = z - F(z)/F'(z) .$$

For example if F is a polynomial of degree d then N is a rational function of degree $\leq d$. Starting with any initial guess z_0 one can form the successive images $z_{k+1} = N(z_k)$. With luck, these will converge towards a solution $\hat{z}$ to the required equation $F(\hat{z}) = 0$.

By definition, $\hat{z}$ is a root of F of *multiplicity* m if the Taylor expansion of F about $\hat{z}$ has the form

$$F(z) = a(z - \hat{z})^m + \text{(higher terms)} ,$$

with $a \neq 0$ and $m \geq 1$. Show that the fixed points of N in the finite plane are precisely the roots of F, and in fact that every root of F is a attracting fixed point for N with multiplier $\lambda = 1 - 1/m$ where m is the multiplicity. Thus every *simple* root of F, with $m = 1$, is a superattracting fixed point of N, while every root of higher multiplicity is a geometrically attracting fixed point. On the other hand, if F is a polynomial of degree $d > 1$, show that ∞ is the unique repelling fixed point of N, with multiplier $d/(d-1)$. Show that $N(z) = \infty$ for $z \in \mathbb{C}$ only if $F'(z) = 0$, $F(z) \neq 0$. Show that N has derivative

$$N'(z) = \frac{F(z)F''(z)}{F'(z)^2} .$$

Problem 4-h. Liapunov stability. A point $z_0 \in \hat{\mathbb{C}}$ is *stable in the sense of Liapunov* for a rational map f if the orbit of any point which is sufficiently close to z_0 remains uniformly close to the orbit of z_0 for all time. More precisely, for every $\epsilon > 0$ there should exist $\delta > 0$ so that if z has spherical distance $\sigma(z_0, z) < \delta$ then $\sigma(f^{\circ n} z_0, f^{\circ n} z) < \epsilon$ for all n. Show that a point is Liapunov stable if and only if it belongs to the Fatou set.

Problem 4-i. Fatou components. If Ω is a connected component of the Fatou set of f, show that $f(\Omega)$ is also a connected component of Fatou(f).

Problem 4-j. Baire's Theorem and transitivity. For any locally compact space X, prove Baire's Theorem that any countable intersection $U_1 \cap U_2 \cap \cdots$ of dense open subsets of X is again dense. (Within any non-vacuous open set $V \subset X$ choose a nested sequence

$$K_1 \supset K_2 \supset K_3 \supset \cdots$$

of compact sets $K_j \subset U_j$ with non-vacuous interior, and take the intersection.)

A map $f : X \to X$ is called *topologically transitive* if for every pair U and V of non-vacuous open subsets there exists an integer $n \geq 0$ so that $f^{\circ n}(U) \cap V$ is non-vacuous. (Compare 4.7.) If this condition is satisfied, and if there is a countable basis for the open subsets of the locally compact space X, show that a generic orbit under f is dense. (Compare 4.13.)

§5. Dynamics on Hyperbolic Surfaces

This section will begin the discussion of dynamics on Riemann surfaces other than the Riemann sphere. It turns out that the possibilities for dynamics on a hyperbolic surface are rather limited. Let us first recall the definition, emphasizing the possibility that a hyperbolic Riemann surface S may well be non-compact.

For a holomorphic map $f : S \to S$ of an arbitrary Riemann surface, the *Fatou set* of f is the union of all open sets $U \subset S$ such that every sequence of iterates $f^{\circ n_j}|_U$ either

(1) contains a locally uniformly convergent subsequence, or

(2) contains a subsequence which diverges locally uniformly to infinity in S, so that the images of a compact subset of U eventually leave any compact subset of S.

(If S is compact, then no sequence can diverge to infinity, so that Case (2) can never occur.) As usual, the complement of the Fatou set is called the *Julia set*.

Remark. In the special case of a surface S which can be described as an open subset with compact closure within a larger Riemann surface T, a completely equivalent condition would be the following. *A point $z \in S$ belongs to the Fatou set of f if and only if, for some neighborhood U of z, any sequence of iterates $f^{\circ n}|_U$ considered as maps from U to T contains a subsequence which converges locally uniformly to a map $U \to T$.* (Compare 3.6. If the limit map does not take values in S, then it is necessarily constant.)

> **Lemma 5.1. No Julia set.** *For any map $f : S \to S$ of a hyperbolic surface, the Julia set $J(f)$ is vacuous. In particular, f can have no repelling points, parabolic points, or basin boundaries.*

Proof. This follows immediately from Corollary 3.3, together with the proofs of 4.3, 4.4, and 4.9. □

In fact we can give a much more precise statement, essentially due to Fatou. (Compare §16.)

> **Classification Theorem 5.2.** *For any holomorphic map $f :$ $S \to S$ of a hyperbolic Riemann surface, exactly one of the following four possibilities holds:*
>
> • **Attracting Case.** *If f has an attracting fixed point, then it follows from 5.1 (or from Problem 1-h) that all orbits under f*

converge towards this fixed point. The convergence is uniform on compact subsets of S.

- **Escape.** *If some orbit under f has no accumulation point in S, then no orbit has an accumulation point. In fact, for any compact set $K \subset S$ there exists an integer n_K so that $K \cap f^{\circ n}(K) = \emptyset$ for $n \geq n_K$.*

- **Finite Order.** *If f has two distinct periodic points, then some iterate $f^{\circ n}$ is the identity map and every point of S is periodic.*

- **Irrational Rotation.** *In all other cases, (S, f) is a rotation domain. That is, S is isomorphic either to a disk $\mathbb{D}$, to a punctured disk $\mathbb{D} \smallsetminus \{0\}$, or to an annulus*

$$\mathbb{A}_r = \{z : 1 < |z| < r\},$$

and f corresponds to an irrational rotation, $z \mapsto e^{2\pi i \alpha} z$ with $\alpha \notin \mathbb{Q}$.

Much later, in §16, we will apply this Theorem to the case where S is an open subset of the Riemann sphere and f is a rational map carrying this set into itself. Here is an important example.

Corollary 5.3. Siegel disks. *Let f be a rational map of degree $d \geq 2$. If a connected component U of the Fatou set $\widehat{\mathbb{C}} \smallsetminus J$ contains an indifferent fixed point,*

$$f(z_0) = z_0, \quad |f'(z_0)| = 1,$$

then U is conformally isomorphic to the unit disk $\mathbb{D}$ in such a way that $f|_U$ corresponds to an irrational rotation of the disk.

The proof, assuming 5.2, is immediate, since U is clearly hyperbolic and maps into itself under f. (No iterate $f^{\circ n}$ can be the identity map, since f has degree ≥ 2.) The actual existence of non-linear rational maps which possess such a rotation domain U is highly non-trivial, and will be discussed in §11.

In the special case where S is equal to the open unit disk $\mathbb{D}$, the following more precise statement was proved by Denjoy in 1926, refining an earlier result by Wolff.

Denjoy-Wolff Theorem 5.4. *Let $f : \mathbb{D} \to \mathbb{D}$ be any holomorphic map. Then either:*

(1) f is a "rotation" (with respect to the Poincaré metric) about some fixed point $z_0 \in \mathbb{D}$, or else

(2) *the successive iterates $f^{\circ n}$ converge, uniformly on compact subsets of $\mathbb{D}$, to a constant function $z \mapsto c_0$, where c_0 may belong either to the open disk $\mathbb{D}$ or to the boundary circle $\partial \mathbb{D}$.*

(This is sharper than 5.2 only in the Escape Case: If some orbit has no accumulation point in $\mathbb{D}$, then every orbit must converge to a single boundary point of $\mathbb{D}$.) According to 1.8, there is an automorphism of $\widehat{\mathbb{C}}$ carrying $\mathbb{D}$ to the upper half-plane $\mathbb{H}$, so the analogue of 5.3 is true for $\mathbb{H}$ also. Here are three examples: If $f : \mathbb{H} \to \mathbb{H}$ is either the parabolic automorphism $z \mapsto z + 1$ or the hyperbolic automorphism $z \mapsto 2z$ or the embedding $z \mapsto z + i$, then all orbits in $\mathbb{H}$ converge within $\widehat{\mathbb{C}}$ to the single boundary point ∞.

Theorem 5.4 is actually not very useful for our purposes, since it is no longer true if we replace $\mathbb{D}$ or $\mathbb{H}$ by an arbitrary hyperbolic open subset of $\widehat{\mathbb{C}}$. (See Problem 5-a.) However, we can make the following statement, which will be useful.

Lemma 5.5. Convergence to a boundary fixed point.
Suppose that U is a hyperbolic open subset of a compact Riemann surface, and that the map $f : U \to U$ extends continuously to the boundary ∂U, with at most a finite number of fixed points in ∂U. If some orbit of f in U has no accumulation point within U, then all orbits in U must converge within the closure $\overline{U}$ to a single boundary fixed point

$$\hat{z} = f(\hat{z}) \in \partial U.$$

This convergence is uniform on compact subsets of U.

The proofs follow.

Proof of 5.2. Choose some base point $p_0 \in S$ and consider the orbit $p_0 \mapsto p_1 \mapsto p_2 \mapsto \cdots$ under f. It may happen that this orbit diverges to infinity, so that the Poincaré distance satisfies

$$\lim_{n \to \infty} \operatorname{dist}(p_n, p_0) = \infty.$$

If this happens, then for any point q_0 within the ball of radius r about p_0, the corresponding orbit $q_0 \mapsto q_1 \mapsto \cdots$ satisfies $\operatorname{dist}(q_n, p_n) \leq r$, and hence

$$\operatorname{dist}(q_n, p_0) \geq \operatorname{dist}(p_n, p_0) - r \to \infty.$$

Thus all orbits diverge to infinity in S, and this divergence is uniform on compact subsets of S.

Otherwise, if $\text{dist}(p_n, p_0)$ does not tend to infinity, then we can find infinitely many p_n within some bounded neighborhood of p_0. These must have some accumulation point $\hat{p} \in S$. Choose integers $n(1) < n(2) < \cdots$ so that the sequence $\{p_{n(j)}\}$ converges to $\hat{p}$, and consider the sequence of maps $g_j = f^{\circ(n(j+1)-n(j))}$. Then g_j maps $p_{n(j)}$ to $p_{n(j+1)}$. If r_j is the Poincaré distance between $\hat{p}$ and $p_{n(j)}$, it follows that $\text{dist}(g_j(\hat{p}), p_{n(j+1)}) \le r_j$, hence

$$\text{dist}(g_j(\hat{p}), \hat{p}) \le r_j + r_{j+1} \qquad (5:1)$$

by the triangle inequality. Let r be the maximum of $\{r_j\}$. (This exists since the r_j converge to zero.) It follows that the points $g_j(\hat{p})$ all lie within some compact ball $B_{2r} \subset S$. Hence by the Hyperbolic Compactness Theorem 3.7, it follows that the maps g_j all lie within a compact subset of the space $\text{Hol}(S, S)$. Therefore we can choose an accumulation point g of $\{g_j\}$ within $\text{Hol}(S, S)$. Furthermore, since $r_j + r_{j+1} \to 0$ as $j \to \infty$, it follows from (5:1) that $g(\hat{p}) = \hat{p}$.

Distance Decreasing Case. If f decreases Poincaré distances, then every iterate of f must satisfy

$$\text{dist}(f^{\circ n}(p), f^{\circ n}(q)) \le \text{dist}(f(p), f(q)) < \text{dist}(p, q)$$

for $p \ne q$, hence the limit g must also decrease Poincaré distances. The maps f and g commute, since g is a limit of iterates of f, hence f must map the fixed point $\hat{p} = g(\hat{p})$ to a fixed point $f(\hat{p}) = f(g(\hat{p})) = g(f(\hat{p}))$ of g. But g cannot have two distinct fixed points, since it decreases the Poincaré distance. This proves that $\hat{p} = f(\hat{p})$ is also a fixed point for f. Evidently it is an attracting fixed point, so all orbits under f must converge to $\hat{p}$.

Distance Preserving Case. Now suppose that f is a local isometry for the Poincaré metric. *Then we will show that some sequence of iterates of f converges locally uniformly to the identity map of S.* Proceeding as above, some sequencee of iterates g_j of f converges to a map g which has a fixed point $\hat{p}$. The multiplier at this fixed point must have absolute value equal to 1, say $g'(\hat{p}) = e^{2\pi i \alpha}$. Whether or not the angle α is rational, we can choose some multiple $m\alpha$ which is arbitrarily close to an integer, and conclude that $g^{\circ m}$ has multiplier arbitrarily close to $+1$ at $\hat{p}$. On the other hand, these iterates of g belong to a normal family by 3.7, so we can choose a subsequence $\{g^{m(j)}\}$ which converges locally uniformly throughout S, with multiplier at $\hat{p}$ converging to $+1$. The limit function has multiplier equal to $+1$. Lifting to the universal covering and applying the Schwarz Lemma, we see that this limit is indeed the identity map of

S . Finally, it is not difficult to reduce this double limit of iterates of f to a single limit.

To complete the proof of 5.2, we must prove the following.

Lemma 5.6. Iterates near the identity map. *If $f : S \to S$ is a map of a hyperbolic surface, with the property that some sequence of iterates $f^{\circ m(i)}$ converges locally uniformly to the identity map, then either f has finite order, or else S is isomorphic to $\mathbb{D}$ or $\mathbb{D} \smallsetminus \{0\}$ or to an annulus $\mathbb{A}_r$, and f corresponds to an irrational rotation.*

(A similar assertion holds for non-hyperbolic surfaces: Compare Problem 6-d.)

Proof of 5.6. First note that f must be one-to-one. For if $f(p) = f(q)$ with $p \neq q$, then any limit of iterates of f must also map p and q to the same point, so no such limit can be the identity map. Similarly, note that f must be onto. Suppose to the contrary that $f(S)$ were a proper subset of S , with say $p \notin f(S)$. If B is a closed disk neighborhood of p , then any map g sufficiently close to the identity map of S must map B to a set $g(B)$ containing p . Hence no such g can be an iterate of f . Combining these two statements, we see that f must be a conformal automorphism of the surface S .

In the simply connected case, the automorphisms of $S \cong \mathbb{D}$ have been described in Theorem 1.11. (See also Problems 1-e and 2-e.) Evidently the 'hyperbolic' and 'parabolic' automorphisms, with no interior fixed point, behave as in the Escape Case, with no iterate close to the identity map. Thus the only automorphisms satisfying the hypothesis of 5.6 are the rotations about some fixed point.

For the non-simply connected case, we have to work a bit harder. Suppose that the sequence of maps $f^{\circ m(j)}$ converges, uniformly on compact sets, to the identity map of S . Lifting to the universal covering surface, we obtain a sequence of automorphisms $F^{\circ m(j)} : \tilde{S} \to \tilde{S}$ which converge to the identity modulo the action of the group Γ of deck transformations. In other words, given a compact set $K \subset \tilde{S}$, for j sufficiently large we can find a deck transformation γ_j so that the composition $F_j = \gamma_j \circ F^{\circ m(j)}$ is uniformly close to the identity throughout K .

Now note that each F_j induces a group homomorphism $\gamma \mapsto \gamma'$ from Γ to itself satisfying the identity $F_j \circ \gamma = \gamma' \circ F_j$. (See Problem 2-b. Since both $F_j \circ \gamma$ and F_j cover the same map from S to itself, it follows that there is some deck transformation γ' carrying $F_j \circ \gamma(\tilde{p})$ to $F_j(\tilde{p})$, and it is easy to check that γ' does not depend on the choice of $\tilde{p} \in \tilde{S}$.) Therefore

$$\gamma' \; = \; F_j \circ \gamma \circ F_j^{-1} \; .$$

If F_j is very close to the identity, then γ' will be very close to γ throughout some large compact set. But Γ is a discrete group, so this implies that $\gamma' = \gamma$ or in other words

$$F_j \circ \gamma \; = \; \gamma \circ F_j$$

provided that j is sufficiently large. If some F_j is actually equal to the identity map, then some iterate of f is the identity map of S . Let us assume that this is not the case, so that no F_j is the identity map.

Recall from Theorem 1.11 that each non-identity element g in the automorphism group $\mathcal{G}(\tilde{S}) \cong \mathcal{G}(\mathbb{D})$ belongs to a unique maximal commutative subgroup, which we will denote by $\mathcal{C}(g)$. Thus two non-identity elements g_1 and g_2 in $\mathcal{G}(\tilde{S})$ commute if and only if $\mathcal{C}(g_1) = \mathcal{C}(g_2)$. In particular, any non-identity $\gamma \in \Gamma \subset \mathcal{G}(\tilde{S})$ determines such a group $\mathcal{C}(\gamma)$, and any F_j which is sufficiently close to the identity map must satisfy $\mathcal{C}(F_j) = \mathcal{C}(\gamma)$. But the same is true for any other non-identity element of Γ . This proves that the commutative group $\mathcal{C}(\gamma) \subset \mathcal{G}(\tilde{S})$ is independent of the particular choice of γ . We will denote this group briefly by $\mathcal{C}(\Gamma)$.

In particular, it follows that Γ is a commutative group. Since we have assumed that S is not simply connected, this implies that S must be either an annulus or a punctured disk. (Problem 2-g.) Furthermore, if j is large then $F_j = \gamma_j \circ F^{om(j)}$ belongs to (Γ) , hence $F^{om(j)}$ does also. But F commutes with $F^{om(j)}$, so F also belongs to $\mathcal{C}(\Gamma)$.

If the one parameter group $\mathcal{C}(\Gamma) \subset \mathcal{G}(\tilde{S})$ consists of parabolic transformation, then it will be convenient to use the upper half-plane model $\tilde{S} \cong \mathbb{H}$, and to identify $\mathcal{C}(\Gamma)$ with the group of real translations $w \mapsto w + c$. On the other hand, if $\mathcal{C}(\Gamma)$ consists of hyperbolic transformations, then it is convenient to use the infinite band model, as in Problem 2-f. In this case also, we can identify $\mathcal{C}(\Gamma)$ with the group of real translations $w \mapsto w + c$. In either case, the non-trivial discrete subgroup Γ must be cyclic, generated by some translation $w \mapsto w + c_0$, and the map F must correspond to some other translation $w \mapsto w + c'$. Now setting $z = e^{2\pi i w / c_0}$ we see that F corresponds to a rotation of an annulus or punctured disk, as required. This completes the proof of 5.6 and 5.2 $\quad \square$

Proof of the Denjoy-Wolff Theorem 5.4. Let $f : \mathbb{D} \to \mathbb{D}$ be any holomorphic map. The following argument is taken from a lecture of Beardon, as communicated to me by Shishikura. For any $\epsilon > 0$, let us approximate f by the map $f_\epsilon(z) = (1 - \epsilon)f(z)$ from $\mathbb{D}$ into a proper subset of itself. Each map f_ϵ has a unique fixed point z_ϵ. (Compare Problem 2-j, or note that existence of a fixed point follows from the Brouwer Fixed Point Theorem applied to the disk $(1 - \epsilon)\overline{\mathbb{D}}$, and that uniqueness is clear since f_ϵ decreases Poincaré distances.) Since the closed disk $\overline{\mathbb{D}}$ is compact, we can choose a sequence $\{\epsilon_i\}$ tending to zero so that the corresponding fixed points z_{ϵ_i} converge to some limit $\hat{z} \in \overline{\mathbb{D}}$. If $|\hat{z}| < 1$, then $\hat{z}$ is a fixed point of f, and the conclusion follows easily from 5.2. Assume then that $\hat{z} \in \partial\mathbb{D}$. Choose some arbitrary base point $z_0 \in \mathbb{D}$, and let r_i be the Poincaré distance between z_0 and z_{ϵ_i}. Let B_i be the closed neighborhood of Poincaré radius r_i which is centered at z_{ϵ_i} and has the base point z_0 on its boundary. Since the map f_{ϵ_i} reduces Poincaré distances, it necessarily carries B_i into itself. These neighborhoods B_i are actually round disks with respect to the Euclidean metric also. (However the Euclidean center is usually different from the Poincaré center. Compare Problem 2-c.) As $i \to \infty$, the round disks B_i must tend to a limit B_∞, which can only be the round disk which is tangent to the unit circle at $\hat{z}$ and whose boundary passes through z_0. (By definition, such a disk tangent to the circle at infinity of $\mathbb{D}$ is called a "horodisk" in $\mathbb{D}$.) It follows by continuity that f maps $\mathbb{D} \cap B_\infty$ into itself. In particular, it follows that the entire orbit of z_0 under f must be contained in B_∞. On the other hand, we know by 5.2 that the orbit of z_0 must tend to the boundary of $\mathbb{D}$. But a sequence in B_∞ which tends to the boundary of $\mathbb{D}$ can only tend to the point of tangency $\hat{z}$. It follows easily that all orbits in $\mathbb{D}$ tend to the same limiting point $\hat{z}$, as required. □

Proof of Lemma 5.5. Let U be a hyperbolic open subset of the Riemann surface S. Suppose that $f : \overline{U} \to \overline{U}$ is continuous on the compact set $\overline{U}$, and maps U holomorphically into itself, and suppose that some orbit $p_0 \mapsto p_1 \mapsto p_2 \mapsto \cdots$ in U has no accumulation point in U. It follows that the Poincaré distance $\text{dist}_U(p_0, p_n)$ must tend to infinity as $n \to \infty$. Choose some continuous path $p : [0, 1] \to U$ from the point $p_0 = p(0)$ to $f(p_0) = p(1)$, and continue this path inductively for all $t \geq 0$ by setting $p(t + 1) = f(p(t))$. Let δ be the diameter of the image $p[0, 1]$ in the Poincaré metric for U. Then each successive image $p[n, n + 1]$ must also have diameter $\leq \delta$. It follows that $\text{dist}_U(p_0, p(t))$ also tends to infinity as $t \to \infty$.

Let $\hat{p}$ be any accumulation point of $\{p(t)\}$ in ∂U as $t \to \infty$. It follows from 3.4 that, for any neighborhood V of $\hat{p}$ we can find a smaller neighborhood W, so that any set of Poincaré diameter δ which intersects $U \cap W$ must be contained in V. Hence, for any such V, we can find images $p[n, n + 1]$ which are contained in V. Since f maps $p(n)$ to

$p(n+1)$, it follows by continuity that $f(\hat{p}) = \hat{p}$. *Thus every accumulation point of the path* $p : [0, \infty) \to U$ *in* ∂U *must be a fixed point of* f. On the other hand, it is not difficult to show that the set of all accumulation points of $p(t)$ as $t \to \infty$ is a connected set. (Problem 5-b.)

Now assume that f has only finitely many fixed points in ∂U. Since a finite connected set can only be a single point, it follows that $p(t)$ converges to a single point $\hat{p} \in \partial U$ as $t \to \infty$. In particular, the orbit $p_0 \mapsto p_1 \mapsto \cdots$ converges to $\hat{p}$. Now consider an arbitrary orbit $q_0 \mapsto q_1 \mapsto \cdots$ under f, if $\mathrm{dist}_U(p_0, q_0) = r$ then $\mathrm{dist}_U(p_n, q_n) \le r$. Using 3.4, it follows that the sequence $\{q_n\}$ also converges to $\hat{p}$, and it is easy to check that this convergence is uniform on compact subsets of U □

Problem 5-a. A badly behaved example. Given numbers $1 > a_1 > a_2 > \cdots$ converging to 0, let $U \subset \mathbb{C}$ be obtained from the open unit square $(0, 1) \times (0, 1)$ by removing the line

$$[a_n, \quad 1] \times \{a_n\} \quad \text{for each odd value of } n \text{ and removing}$$
$$[0, 1 - a_n] \times \{a_n\} \quad \text{for each even value of } n,$$

as illustrated. Given a base point z_0 in the open set U, and given a boundary point $\hat{z} \in \partial U$ which is *not* on the bottom edge $[0, 1] \times \{0\}$, it can be shown that there is at least one geodesic ray $p : [0, \infty) \to U$ with respect to the Poincaré metric on U, parametrized by arclength, such that $z_0 = p(0)$ and $\hat{z} = \lim_{t \to \infty} p(t)$. (Compare 17.9 below.) Assuming this statement, show that there must exist another geodesic ray starting at z_0 so that the set of accumulation points of $p(t)$ as $t \to \infty$ is the entire bottom edge $[0, 1] \times \{0\}$. According to Problem 2-e, there is a conformal automorphism $f : U \to U$ such that $f(p(t)) = p(t + 1)$. Show that the set of accumulation points for the orbit $z_0 \mapsto z_1 \mapsto \cdots$ is this same bottom edge.

Problem 5-b. Accumulation points of a path. In any Hausdorff space X, show that the closure of a connected set is connected, and show that the intersection of any nested sequence $K_1 \supset K_2 \supset \cdots$ of compact connected sets is again connected. Now consider an infinite path $p : [0, \infty) \to X$ in a compact Hausdorff space. Show that the set of all accumulation points of $p(t)$ as $t \to \infty$ can be identified with the intersection of closures

$$\bigcap_t \overline{p[t, \infty)} \, ,$$

and therefore is a non-vacuous compact connected set.

§6. Dynamics on Euclidean Surfaces

This section considers surfaces S such that the universal covering $\hat{S}$ is conformally isomorphic to the complex numbers $\mathbb{C}$. Thus S can be either $\mathbb{C}$ itself, or the punctured plane $\mathbb{C} \smallsetminus \{0\} \cong \mathbb{C}/\mathbb{Z}$, or a torus $\mathbb{T} = \mathbb{C}/\Lambda$. (Compare §2.) It turns out that the torus case is interesting but quite easy to understand, while the remaining two cases are extremely difficult.

In the case of a torus $\mathbb{T} = \mathbb{C}/\Lambda$, where Λ is a lattice in $\mathbb{C}$, we will prove the following.

6.1. Theorem. *Every holomorphic map $f : \mathbb{T} \to \mathbb{T}$ is an affine map, $f(z) \equiv \alpha z + c \pmod{\Lambda}$, with degree $d = |\alpha|^2$. The corresponding Julia set $J(f)$ is either the empty set or the entire torus according as $d \leq 1$ or $d > 1$.*

If $\alpha \neq 1$, note that f has a fixed point $z_0 = c/(1 - \alpha)$, and hence is conjugate to the linear map

$$z \mapsto f(z + z_0) - z_0 = \alpha z.$$

For the description of which coefficients α can occur, see Problem 6-a.

Proof of 6.1. To fix our ideas, suppose that $\mathbb{T} = \mathbb{C}/\Lambda$ where the lattice $\Lambda \subset \mathbb{C}$ is spanned by the two numbers 1 and τ, and where $\tau \notin \mathbb{R}$. Any holomorphic map $f : \mathbb{T} \to \mathbb{T}$ lifts to a holomorphic map $F : \mathbb{C} \to \mathbb{C}$ on the universal covering surface. Note first that there exists a lattice element $\alpha \in \Lambda$ so that

$$F(z + 1) = F(z) + \alpha \qquad \text{for all} \qquad z \in \mathbb{C}.$$

For we certainly have $F(z + 1) \equiv F(z) \pmod{\Lambda}$, and the difference function $F(z+1) - F(z) \in \Lambda$ must be constant since $\mathbb{C}$ is connected and the target space Λ is discrete. Similarly, there exists β so that $F(z+\tau) = F(z) + \beta$ for all z. Now let $g(z) = F(z) - \alpha z$, so that $g(z + 1) = g(z)$. Then

$$g(z + \tau) = F(z + \tau) - \alpha(z + \tau) = g(z) + (\beta - \alpha\tau).$$

We claim that g must be constant, say $g(z) = c$ for all z. In fact g gives rise to a map from $\mathbb{T}$ to the quotient space $\mathbb{C}/(\beta - \alpha\tau)\mathbb{Z}$, which is either $\mathbb{C}$ itself or an infinite cylinder according as $\beta - \alpha\tau$ is zero or non-zero. In either case, this quotient is a non-compact Riemann surface, while $\mathbb{T}$ is compact, so such a map must be constant by the Maximum Modulus Principle. Now $F(z) = g(z) + \alpha z = \alpha z + c$, as required. Since this map multiplies areas by $|\alpha|^2$, it follows easily that f has degree equal to $|\alpha|^2$.

Further properties of f depend on the multiplier α. If $|\alpha| \leq 1$, then the derivatives $|d f^{on}(z)/dz| = |\alpha^n|$ are uniformly bounded, so the domain of normality for $\{f^{on}\}$ is the entire torus $\mathbb{T}$. In other words, the Fatou set of f is equal to $\mathbb{T}$. On the other hand, if $|\alpha| > 1$, then $|d f^{on}(z)/dz| = |\alpha^n| \to \infty$ as $n \to \infty$, and it follows that the Julia set of f is the entire torus. $\square$

For further information, see Problems 6-a, b below.

Figure 5. The function $z \mapsto \sin(z)$ can be considered as a holomorphic map from the cylinder $\mathbb{C}/2\pi\mathbb{Z}$ to itself. In this case the Julia set, shown in black, has infinite area (McMullen). The Fatou set $(\mathbb{C}/2\pi\mathbb{Z}) \smallsetminus J$ is dense, but conjecturally has finite area. (The region shown is $[-.5, \pi + .5] \times [-1, 4]$.)

The non-compact Euclidean surfaces. Recall that there are only two non-compact surfaces covered by $\mathbb{C}$. First suppose that S is the complex plane $\mathbb{C}$ itself. We can distinguish two different classes of holomorphic maps $\mathbb{C} \to \mathbb{C}$. A *polynomial map* of $\mathbb{C}$ extends uniquely over the Riemann sphere $\hat{\mathbb{C}}$. Hence the theory of polynomial mappings can be subsumed as a special case of the theory of rational maps of $\hat{\mathbb{C}}$. (Compare §9 and §18.) On the other hand, *transcendental mappings* from $\mathbb{C}$ to itself form an essential distinct and more difficult subject of study. Such mappings have been studied for more than sixty years by many authors, starting with Fatou himself. (See especially Baker, [Ba1, Ba2].) Even iteration of the exponential map $\exp : \mathbb{C} \to \mathbb{C}$ provides a number of quite challenging problems.

For example, according to Lyubich (1987) and Rees (1986), for Lebesgue almost every starting point $z \in \mathbb{C}$, the set of accumulation points for the orbit of z is equal to the orbit $\{0, 1, e, e^e, \ldots\}$ of zero. (This assertion is an amusing subject for computer experimentation: Random empirical orbits seem to land exactly at 0 after relatively few iterations, unless they first encounter an overflow error.) However, according to Misiurewicz (1981) the Julia set of the exponential map is the entire complex plane, hence a generic orbit is everywhere dense in the plane. (Compare 4.13. A proof that $J(\exp) = \mathbb{C}$ is included in Devaney (1989). For a map of the interval with the analogous property that a generic orbit is dense but almost every orbit is not, see Bruin, Keller, Nowicki and van Strien.)

Further information about iterated transcendental functions may be found for example in Devaney (1986), Goldberg & Keen, and in Eremenko & Lyubich (1990), (1992). The study of iterated maps from the *cylinder* $\mathbb{C}/\mathbb{Z} \cong \mathbb{C} \smallsetminus \{0\}$ to itself is closely related, and is also a difficult and interesting subject. See for example Keen (1988). Note that any periodic function from $\mathbb{C}$ to itself can also be considered as a function from the cylinder to itself. (Compare Figure 5.)

Remark. The study of iterated *meromorphic* functions $\mathbb{C} \to \widehat{\mathbb{C}}$, although surely of great interest, does not fit into the framework we have described, since compositions are not everywhere defined. Compare Bergweiler (1993).

We conclude with problems for the reader.

Problem 6-a. The derivative of a torus map. Consider the torus $\mathbb{T} = \mathbb{C}/\Lambda$, where we may assume that $\Lambda = \mathbb{Z} \oplus \tau\mathbb{Z}$ with $\tau \notin \mathbb{R}$. Given $\alpha \in \mathbb{C}$, show that there exists a holomorphic map $f(z) \equiv \alpha z + c$ from $\mathbb{T}$ to itself with derivative α if and only if $\alpha\Lambda \subset \Lambda$, or in other words if and only if both α and $\alpha\tau$ belong to Λ. Show that an arbitrary integer $\alpha \in \mathbb{Z}$ will satisfy this condition. On the other hand, show that there exists such a map with derivative $\alpha \notin \mathbb{Z}$ if and only if α satisfies a quadratic equation of the form

$$\alpha^2 + p\alpha + d = 0 ,$$

where $d = |\alpha|^2$ is the degree, and where p is an integer with $p^2 < 4d$. (Such a torus is said to admit "*complex multiplications*".) For a map of degree $d = |\alpha|^2 = 1$ show that α must be an m-th root of unity with $m = 1, 2, 3, 4$, or 6. If $m \neq 1$, conclude that $f^{\circ m}$ must be the identity map. Show that the cases $m = 3, 4, 6$ occur for suitably chosen lattices, and that the cases $m = 1, 2$ occur for an arbitrary lattice. In the special case $\alpha = 1$, show that the closure of every orbit under f is either a finite set, a finite union of parallel circles, or the full torus $\mathbb{T}$.

Problem 6-b. Periodic points of torus maps. If $\alpha \neq 0$, show that any equation of the form $f(z) = z_0$ has exactly $d = |\alpha|^2$ solutions $z \in \mathbb{T}$. If $\alpha \neq 1$, show that f has exactly $|\alpha - 1|^2$ fixed points. (In particular, both $|\alpha|^2$ and $|\alpha-1|^2$ are necessarily integers.) More generally, if $|\alpha| > 1$ show that the equation $f^{\circ n}(z) = z$ has exactly $|\alpha^n - 1|^2$ solutions in $\mathbb{T}$, all repelling with multiplier α^n. Show that the periodic points of f are everywhere dense in $\mathbb{T}$ whenever $\alpha \neq 0, 1$.

Problem 6-c. Grand orbit finite points. Show that a nonlinear holomorphic map $f : \mathbb{C} \to \mathbb{C}$ has at most one grand orbit finite point. Show by examples such as $f(z) = \lambda z e^z$ and $f(z) = z^2 e^z$ that this fixed point need not be attracting, and in fact can have arbitrary multiplier.

Problem 6-d. Non-hyperbolic rotation domains. Prove the following analogue of Lemma 5.6. If $f : S \to S$ is a self-map of the non-hyperbolic surface S such that some sequence of iterates of f converges locally uniformly to the identity map, but no iterate is actually equal to the identity, show that f is either a rotation of $\widehat{\mathbb{C}}$ or $\mathbb{C}$ or $\mathbb{C} \smallsetminus \{0\}$, or a translation of a torus, up to conformal isomorphism.

§7. Smooth Julia Sets

Most Julia sets tend to be complicated fractal subsets of $\widehat{\mathbb{C}}$. However there are three exceptions: According to a Theorem of Hamilton (1995), every Julia set which is a one-dimensional topological manifold must either be a circle or closed line segment up to Möbius automorphism, or must have Hausdorff dimension strictly greater than one. If we count the entire Riemann sphere as another smooth example, it follows that, up to automorphism, there are only three possibilities for smooth Julia sets of rational functions of degree ≥ 2. However, each of these can appear as Julia set for many different rational functions: a property which is itself exceptional. This section will discuss these examples.

Example 1: The Circle. The unit circle appears as Julia set for the mapping $z \mapsto z^{\pm n}$ for any $n \geq 2$. (Compare the discussion of the squaring map in §4.) Other rational maps with this same Julia set are described in Problem 7-b. Similarly, the real axis $\mathbb{R} \cup \infty$, as the image of the unit circle under a conformal automorphism, can appear as a Julia set. (Problem 7-a.)

Example 2: The Interval. Following Ulam and Von Neumann, consider the map $f(z) = z^2 - 2$, which carries the closed interval $I = [-2, 2]$ onto itself. This map, and its generalizations to higher degree, are also known as *Chebyshev polynomials*. (Problem 7-c. For further examples see 7-d.)

> **Lemma 7.1.** *The Julia set J for $f(z) = z^2 - 2$ is equal to the interval $I = [-2, 2]$, and every point outside of I belongs to the attractive basin $\mathcal{A}(\infty)$ of the point at infinity.*

First Proof. For $z_0 \in I$, it is easy to check that both solutions of the equation $f(z) = z_0$ belong to this interval I. Since I contains a repelling fixed point $z = 2$, it follows from Theorem 4.10 that I contains the entire Julia set $J(f)$. On the other hand, the basin $\mathcal{A}(\infty)$ is a neighborhood of infinity whose boundary $\partial \mathcal{A}(\infty)$ is contained in $J(f) \subset I$ by Theorem 4.9. Hence everything outside of $J(f)$ must belong to this basin, or in other words must have orbit escaping to infinity. Since every point of I has bounded orbit, it follows that $J(f) = I$. $\square$

Alternative Proof. We make use of the substitution $g(w) = w + w^{-1}$, which carries the unit circle in a two-to-one manner onto $I = [-2, 2]$. For $z_0 \notin I$, the equation $g(w) = z_0$ has two solutions, one of which lies inside the unit circle and one of which lies outside. Hence g maps the exterior of the closed unit disk isomorphically onto the complement $\mathbb{C} \smallsetminus I$. Since the squaring map in the w-plane is related to f by the identity

$$g(w^2) = g(w)^2 - 2 = f(g(w)),$$

it follows easily that the orbit of z under f either remains bounded or diverges to infinity according as z does or does not belong to this interval. Again using 4.9, it follows that $J(f) = I$. $\square$

Example 3: All of $\widehat{\mathbb{C}}$. The rest of this section will describe a family of examples constructed by S. Lattès, shortly before his death (of typhoid fever) in 1918. Given any lattice $\Lambda \subset \mathbb{C}$ we can form the quotient torus $\mathbb{T} = \mathbb{C}/\Lambda$, as in §2 or §6. Thus $\mathbb{T}$ is a compact Riemann surface, and is also an additive Lie group. Note that the automorphism $z \mapsto -z$ of this surface has just four fixed points. For example, if $\Lambda = \mathbb{Z} + \tau\mathbb{Z}$ is the lattice with basis 1 and τ, where $\tau \notin \mathbb{R}$, then the four fixed points are $0,\ 1/2,\ \tau/2$, and $(1 + \tau)/2$ modulo Λ.

Now form a new Riemann surface S as a quotient of $\mathbb{T}$ by identifying each $z \in \mathbb{T}$ with $-z$. Evidently S inherits the structure of a Riemann surface (although it loses the group structure). In fact we can use $(z - z_j)^2$ as a local uniformizing parameter for S near each of the four fixed points z_j. Thus the natural map $\mathbb{T} \to S$ is two-to-one, except at the four ramification points. To compute the genus of S, we use the following.

7.2. Riemann-Hurwitz Formula. *Let $T \to S$ be a branched covering map from one compact Riemann surface onto another. Then the number of branch points, counted with multiplicity, is equal to $\chi(S)d - \chi(T)$, where χ is the Euler characteristic and d is the degree.*

Sketch of Proof. Choose some triangulation of S which includes all critical values (that is all ramification points) as vertices; and let $a_n(S)$ be the number of n-simplexes, so that $\chi(S) = a_2(S) - a_1(S) + a_0(S)$. In general, each simplex of S lifts to d distinct simplices in T. However, if v is a critical value, then there are too few pre-images of v. The number of missing pre-images is precisely the number of ramification points over v, each counted with an appropriate multiplicity; and the conclusion follows. $\square$

Remark. This proof works also for Riemann surfaces with smooth boundary. The Formula remains true for proper maps between non-compact Riemann surfaces, as can be verified, for example, by means of a direct limit argument.

In our example, since T is a torus $\mathbb{T}$, with Euler characteristic $\chi(\mathbb{T}) = 0$, and since there are exactly four simple branch points, we conclude that $2\chi(S) - \chi(\mathbb{T}) = 4$, or $\chi(S) = 2$. *Using the standard formula $\chi = 2 - 2g$, we conclude that S is a surface of genus zero, isomorphic to the Riemann sphere.* (Remark: The projection map from $\mathbb{T}$ to the sphere $\widehat{\mathbb{C}}$, suitably normalized, is known as the *Weierstrass ℘-function.*)

Now consider the doubling map $z \mapsto 2z$ on $\mathbb{T}$. This commutes with multiplication by -1, and hence induces a map $f : S \to S$. Since the doubling map has degree four, it follows that f is a rational map of degree four. (More generally, in place of the doubling map, we could use any linear map which carries the lattice Λ into itself, as in Problem 6-a.)

7.3. Lattès Theorem. *The Julia set for this rational map f is the entire sphere S.*

Proof. Evidently the doubling map on $\mathbb{T}$ has the property that periodic points are everywhere dense. For example, if r and s are any rational numbers with odd denominator, then $r + s\tau$ is periodic. These periodic orbits are all repelling, since the multiplier is a power of two. Evidently f inherits the same property, and the conclusion follows by 4.3. (Alternatively, given a small open set $U \subset S$, it is not difficult to show that $f^{\circ n}(U)$ is equal to the entire sphere S whenever n is sufficiently large. Hence no sequence of iterates of f can converge to a limit on any open set.) □

In order to pin down just which rational map f has these properties, we must first label the points of S. The four branch points on $\mathbb{T}$ map to four "ramification points" on S, which will play a special role. Let us choose a conformal isomorphism from S onto $\widehat{\mathbb{C}}$ which maps the first three of these points to $\infty, 0, 1$ respectively. The fourth ramification point must then map to some $a \in \mathbb{C} \smallsetminus \{0, 1\}$. In this way we construct a projection map $\wp : T \to \widehat{\mathbb{C}}$ of degree two, which satisfies $\wp(-z) = \wp(z)$, and which has critical values

$$\wp(0) = \infty, \quad \wp(1/2) = 0, \quad \wp(\tau/2) = 1, \quad \wp((1+\tau)/2) = a.$$

(Note: This $\wp$ is a linear function of the usual Weierstrass $\wp$-function.) Here a can be any number distinct from $0, 1, \infty$. In fact, given $a \in \mathbb{C} \smallsetminus \{0, 1\}$, it is not difficult to show that there is one and only one branched covering $\mathbb{T}' \to \widehat{\mathbb{C}}$ of degree two with precisely $\{\infty, 0, 1, a\}$ as ramification points. (Compare Appendix E.) The Riemann-Hurwitz formula shows that this branched covering space $\mathbb{T}'$ is a torus, necessarily isomorphic to $\mathbb{C}/(\mathbb{Z} + \tau\mathbb{Z})$ for some $\tau \notin \mathbb{R}$. The unique deck transformation which interchanges the two pre-images of any point must preserve the linear structure, and hence must be multiplication by -1.

Now the doubling map on $\mathbb{T}$ corresponds under $\wp$ to a specific rational map $f_a : \widehat{\mathbb{C}} \to \widehat{\mathbb{C}}$, where

$$f_a(\wp(z)) = \wp(2z),$$

and where $J(f_a) = \widehat{\mathbb{C}}$ by 7.3. A precise computation of this map f_a is described in Problem 7-g below.

Remark. Mary Rees (1984, 1986a) has proved the existence of many more rational maps with $J(f) = \widehat{\mathbb{C}}$. See also Herman (1984). For any degree $d \geq 2$, let $\mathrm{Rat}(d)$ be the complex manifold consisting of all rational maps of degree d. Rees shows that there is a subset of $\mathrm{Rat}(d)$ of positive measure consisting of maps f which are "ergodic". By definition, this means that any measurable subset of $\widehat{\mathbb{C}}$ which is fully invariant under f must either have full measure or measure zero. It can be shown that any ergodic map must necessarily have $J(f) = \widehat{\mathbb{C}}$.

We will study some of these maps with smooth Julia sets further in 19.9.

Concluding problems.

Problem 7-a. A Newton's method example (Schröder 1871, Cayley 1879). Let $f(z) = z^2 + 1$. If we try to solve the equation $f(z) = 0$ by Newton's method (Problem 4-g), show that we are led to the rational map

$$N(z) = z - f(z)/f'(z) = \tfrac{1}{2}(z - 1/z)$$

from $\widehat{\mathbb{C}} = \mathbb{C} \cup \infty$ to itself. Show that every orbit of N in the upper half-plane converges to $+i$, and that every orbit in the lower half-plane converges to $-i$. Conclude that the Julia set $J(N)$ is equal to $\mathbb{R} \cup \infty$. (Alternatively, note that N is conjugate to $z \mapsto z^2$ under a holomorphic change of coordinates.) More generally, for any quadratic equation with distinct roots, show that $J(N)$ is a straight line together with the point at infinity. What happens for a quadratic equation with double root?

Problem 7-b. Blaschke products. For any $a \in \mathbb{D}$ the map

$$\phi_a(z) = (z - a)/(1 - \bar{a}z)$$

carries the unit disk $\mathbb{D}$ isomorphically onto itself. (Compare 1.7) A finite product of the form

$$f(z) = e^{i\theta}\phi_{a_1}(z)\phi_{a_2}(z)\cdots\phi_{a_n}(z)$$

with $a_j \in D$ is called a *Blaschke product* of degree n. Show that every such f is a rational map which carries $\mathbb{D}$ onto $\mathbb{D}$ and $\widehat{\mathbb{C}} \smallsetminus \overline{\mathbb{D}}$ onto $\widehat{\mathbb{C}} \smallsetminus \overline{\mathbb{D}}$. Conclude that the Julia set $J(f)$ is contained in the unit circle. If $g(z) = 1/f(z)$, interchanging the interior and exterior of the unit circle, show that $J(g)$ is also contained in the unit circle. If $n \geq 2$, and if one of the factors is $\phi_0(z) = z$ show that f has attracting fixed points at zero and infinity, and show that $J(f)$ is the entire unit circle.

Problem 7-c. Chebyshev polynomials. Define monic polynomials
$$P_1(z) = z\,, \quad P_2(z) = z^2 - 2\,, \quad P_3(z) = z^3 - 3z\,, \quad \ldots$$
inductively by the formula $P_{n+1}(z) + P_{n-1}(z) = zP_n(z)$. Show that $P_n(w + w^{-1}) = w^n + w^{-n}$, or equivalently that $P_n(2\cos\theta) = 2\cos(n\theta)$, and show that $P_m \circ P_n = P_{mn}$. For $n \geq 2$ show that the Julia set of $\pm P_n$ is the interval $[-2, 2]$. For $n \geq 3$ show that P_n has $n - 1$ distinct critical points but only two critical values, namely ± 2.

Problem 7-d. More interval Julia sets. Now suppose that f is a Blaschke product with real coefficients, and with an attracting fixed point at the origin. (Compare 7-b.) Show that there is one and only one rational map F of the same degree so that the following diagram is commutative.

$$
\begin{array}{ccc}
\widehat{\mathbb{C}} & \xrightarrow{\ f\ } & \widehat{\mathbb{C}} \\
\downarrow{z+1/z} & & \downarrow{z+1/z} \\
\widehat{\mathbb{C}} & \xrightarrow{\ F\ } & \widehat{\mathbb{C}}
\end{array}
\;,
$$

and show that $J(F) = [-2, 2]$. In the special case $f(z) = z^n$, show that this construction yields the Chebyshev polynomials.

Problem 7-e. Periodic Orbits. Show that the Julia sets studied in Examples 1, 2, 3 have the following extraordinary property. For all but one or two of the periodic orbits $z_0 \mapsto z_1 \mapsto \cdots \mapsto z_n = z_0$, the multiplier $\lambda = f'(z_1) \cdots \cdots f'(z_n)$ satisfies $|\lambda| = d^n$ when J is 1-dimensional, or $|\lambda| = d^{n/2}$ when $J = \widehat{\mathbb{C}}$, where d is the degree. (Compare Problem 19-d.)

Problem 7-f. A quadratic Lattès map. Let $\mathbb{T}$ be the torus $\mathbb{C}/\mathbb{Z}[i]$, where $\mathbb{Z}[i] = \mathbb{Z} \oplus i\mathbb{Z}$ is the lattice of Gaussian integers, and let $L : \mathbb{T} \to \mathbb{T}$ be the linear map $L(z) = (1+i)z$ of degree $|1+i|^2 = 2$. (Compare 6.1 and Problem 6-a.) Let $\wp : \mathbb{T} \to \widehat{\mathbb{C}}$ be the associated Weierstrass map, with $\wp(-z) = \wp(z)$, and let $F = \wp \circ L \circ \wp^{-1}$ be the associated quadratic rational map. Show that F has critical orbits
$$\wp((1+i)/4) \;\mapsto\; \wp(i/2) \;\mapsto\; \wp((1\pm i)/2) \;\mapsto\; \wp(0)$$
and
$$\wp((1-i)/4) \;\mapsto\; \wp(1/2) \;\mapsto\; \wp((1\pm i)/2) \;\mapsto\; \wp(0)\,.$$

Show that the multiplier at the fixed point $\wp(0)$ is equal to $(1+i)^2 = 2i$. After conjugating F by a Möbius automorphism, we may assume that the critical points are ± 1 and the eventual fixed point is at ∞. Show that the most general quadratic map with critical points ± 1 and a fixed point at ∞ has the form $f(z) = a(z + z^{-1}) + b$, and show that the required critical orbit relations are satisfied if and only if $a^2 = -1/2$ and $b = 0$. More precisely, by computing the multiplier of the fixed point at infinity, show that $a = 1/2i$. (Compare Milnor, 1999.)

Problem 7-g. The family of degree four Lattès maps. For the torus $\mathbb{T} = \mathbb{C}/(\mathbb{Z} + \tau\mathbb{Z})$ of Example 3, show that the involution $z \mapsto z + 1/2$ of $\mathbb{T}$ corresponds under $\wp$ to an involution of the form $w \mapsto a/w$ of $\widehat{\mathbb{C}}$, with fixed points $w = \pm\sqrt{a}$. Show that the rational map $f = f_a$ has poles at ∞, 0, 1, a and double zeros at $\pm\sqrt{a}$. Show that f has a fixed point of multiplier $\lambda = 4$ at infinity, and conclude that

$$f(w) = \frac{(w^2 - a)^2}{4w(w - 1)(w - a)}.$$

As an example, if $a = -1$ then

$$f(w) = \frac{(w^2 + 1)^2}{4w(w^2 - 1)}.$$

Show that the correspondence $\tau \mapsto a = a(\tau) \in \mathbb{C} \smallsetminus \{0, 1\}$ satisfies the equations

$$a(\tau + 1) = 1/a(\tau), \quad a(-1/\tau) = 1 - a(\tau),$$

and also $a(-\bar{\tau}) = \bar{a}(\tau)$. Conclude, for example, that $a(i) = 1/2$, and that $a((1 + i)/2) = -1$. (This correspondence $\tau \mapsto a(\tau)$ is an example of an "elliptic modular function", and provides an explicit representation of the upper half-plane $\mathbb{H}$ as a universal covering of the thrice punctured sphere $\mathbb{C} \smallsetminus \{0, 1\}$. Compare Ahlfors (1966) pp. 269-274.)

Problem 7-h. Postcritical finiteness. For each of the six critical points ω of f, show that $f(f(\omega))$ is the repelling fixed point at infinity. (According to 16.5, the fact that each critical orbit terminates on a repelling cycle is already enough to imply that $J(f) = \widehat{\mathbb{C}}$. Note that rational maps satisfying this condition are much more common than Lattès maps.)

LOCAL FIXED POINT THEORY

§8. Geometrically Attracting or Repelling Fixed Points

The next four sections will study the dynamics of a holomorphic map in some small neighborhood of a fixed point. This local theory is a fundamental tool in understanding more global dynamics. It has been studied for well over a hundred years by mathematicians such as Schröder, Kœnigs, Leau, Böttcher, Fatou, Cremer, Siegel, Ecale, Voronin, Cherry, Herman, Bryuno, Yoccoz, and Perez-Marco. In most cases it is now well understood, but a few cases still present extremely difficult problems.

We start by expressing our map in terms of a local uniformizing parameter z, which can be chosen so that the fixed point corresponds to $z = 0$. We can then describe the map by a power series of the form

$$f(z) = \lambda z + a_2 z^2 + a_3 z^3 + \cdots ,$$

which converges for $|z|$ sufficiently small. Recall that the initial coefficient $\lambda = f'(0)$ is called the *multiplier* of the fixed point: it is given a special name since it plays a dominant role in the discussion.

Attracting Points.

Definition. A fixed point p of a map f is *topologically attracting* if it has a neighborhood U so that the successive iterates f^{on} are all defined throughout U, and so that this sequence $\{f^{on}|_U\}$ converges uniformly to the constant map $U \to p$.

Lemma 8.1. Topological characterization of attracting points. *A fixed point for a holomorphic map is topologically attracting if and only if its multiplier satisfies $|\lambda| < 1$.*

Proof. In one direction, this follows from elementary calculus. We can assume as above that the fixed point is $0 = f(0) \in \mathbb{C}$, with Taylor expansion $f(z) = \lambda z + O(z^2)$ as $z \to 0$. In other words there are constants $r_0 > 0$ and C so that

$$|f(z) - \lambda z| \leq C|z^2| \qquad \text{for} \qquad |z| < r_0 . \qquad (8:1)$$

Choose c so that $|\lambda| < c < 1$, and choose $0 < r \leq r_0$ so that $|\lambda| + Cr < c$. For all $|z| < r$, it follows that

$$|f(z)| \leq |\lambda z| + C|z^2| \leq c|z| ,$$

and hence

$$|f^{on}(z)| \leq c^n|z| < c^n r .$$

As $n \to \infty$, this tends uniformly to zero, as required.

Conversely, if f is topologically attracting, then for any sufficiently small disk $\mathbb{D}_\epsilon$ about the origin there exists an iterate $f^{\circ n}$ which maps $\mathbb{D}_\epsilon$ onto a proper subset of itself. By the Schwarz Lemma 1.2, this implies that the multiplier of $f^{\circ n}$ satisfies $|\lambda^n| < 1$, hence $|\lambda| < 1$ as required. $\square$

Definition. An attracting fixed point will be called either *superattracting* or *geometrically attracting*, according as it multiplier is zero, or satisfies $0 < |\lambda| < 1$.

In either case, we will show that f can be reduced to a simple normal form by a suitable change of coordinates. This section considers only the geometrically attracting case $\lambda \neq 0$. In other words, we assume that the origin is not a critical point. The following was proved by G. Kœnigs [*] in 1884.

> **Theorem 8.2. Kœnigs Linearization.** *If the multiplier λ satisfies $|\lambda| \neq 0, 1$, then there exists a local holomorphic change of coordinate $w = \phi(z)$, with $\phi(0) = 0$, so that $\phi \circ f \circ \phi^{-1}$ is the linear map $w \mapsto \lambda w$ for all w in some neighborhood of the origin. Furthermore, ϕ is unique up to multiplication by a non-zero constant.*

In other words, the following diagram is commutative,

$$
\begin{array}{ccc}
U & \xrightarrow{\ f\ } & f(U) \\
\phi \downarrow & & \phi \downarrow \\
\mathbb{C} & \xrightarrow{\ \lambda \cdot\ } & \mathbb{C},
\end{array}
$$

where ϕ is univalent on the neighborhood U of zero. The usefulness of this functional equation

$$\phi \circ f \circ \phi^{-1}(w) \;=\; \lambda w \tag{8:2}$$

had been pointed out some years earlier by E. Schröder. (Compare Alexander.) However, Schröder had been able to find solutions only in very special cases.

Proof of uniqueness. If there were two such maps ϕ and ψ, then the composition

$$\psi \circ \phi^{-1}(w) \;=\; b_1 w + b_2 w^2 + b_3 w^3 + \cdots$$

would commute with the map $w \mapsto \lambda w$. Comparing coefficients of the two resulting power series, we see that $\lambda b_n = b_n \lambda^n$ for all n. Since λ is

---

[*]   Much later Kœnigs became the first Secretary General of the International Mathematical Union where, according to Lehto (1998), he "inflicted damage ... by persisting in maintaining an anti-German policy ...".

neither zero nor a root of unity, this implies that $b_2 = b_3 = \cdots = 0$. Thus $\psi \circ \phi^{-1}(w) = b_1 w$, or in other words $\psi(z) = b_1 \phi(z)$.

Proof of existence when $|\lambda| < 1$. Choose a constant $c < 1$ so that $c^2 < |\lambda| < c$. As in the proof of 8.1, we can choose a neighborhood $\mathbb{D}_r$ of the origin so that $|f(z)| \le c|z|$ for $z \in \mathbb{D}_r$. Thus for any starting point $z_0 \in \mathbb{D}_r$, the orbit $z_0 \mapsto z_1 \mapsto \cdots$ under f converges geometrically towards the origin, with $|z_n| \le rc^n$. By Taylor's Theorem $(8:1)$ we have $|f(z) - \lambda z| \le C|z^2|$ for $z \in \mathbb{D}_r$, hence

$$|z_{n+1} - \lambda z_n| \le C|z_n|^2 \le Cr^2 c^{2n}.$$

Setting $k = Cr^2/|\lambda|$, it follows that the numbers $w_n = z_n/\lambda^n$ satisfy

$$|w_{n+1} - w_n| \le k(c^2/|\lambda|)^n.$$

These differences converge uniformly and geometrically to zero. Thus the holomorphic functions $z_0 \mapsto w_n(z_0)$ converge, uniformly throughout $\mathbb{D}_r$, to a holomorphic limit $\phi(z_0) = \lim_{n \to \infty} z_n/\lambda^n$. (Compare 1.4.) The required identity $\phi(f(z)) = \lambda \phi(z)$ follows immediately. Furthermore, since each correspondence $z_0 \mapsto w_n = z_n/\lambda^n$ has derivative one at the origin, it follows the the limit function ϕ has derivative $\phi'(0) = 1$, and hence is a local conformal isomorphism.

Proof when $|\lambda| > 1$. The statement in this case follows immediately by applying the argument above to the map f^{-1}, which can be defined as a single valued holomorphic function in some neighborhood of zero, with multiplier satisfying $0 < |\lambda^{-1}| < 1$. □

8.3. Remark. More generally, suppose that we consider a family of maps f_α of the form

$$f_\alpha(z) = \lambda(\alpha)z + b_2(\alpha)z^2 + \cdots$$

which depend holomorphically on one (or more) complex parameters α and have multiplier satisfying $|\lambda(\alpha)| \ne 0, 1$. Then a similar argument shows that *the Kœnigs function* $\phi(z) = \phi_\alpha(z)$ *depends holomorphically on* α. (This fact will be important in 11.15.) To prove this statement, first fix some $0 < c < 1$ and suppose that $|\lambda(\alpha)|$ varies through some compact subset of the interval (c^2, c). Then the convergence in the proof of 8.2 is uniform in α. The more general case follows easily. □

For the moment, let us consider only the attracting case $0 < |\lambda| < 1$. We can restate 8.2 in a more global form as follows. Suppose that $f : S \to S$ is a holomorphic map from a Riemann surface into itself with an attracting fixed point $\hat{p} = f(\hat{p})$ of multiplier $\lambda \ne 0$. Recall from §4 that the total *basin of attraction* $\mathcal{A} = \mathcal{A}(\hat{p}) \subset S$ consists of all $p \in S$

for which $\lim_{n\to\infty} f^{\circ n}(p)$ exists and is equal to $\hat{p}$. The *immediate basin* $\mathcal{A}_0$ is defined to be the connected component of $\mathcal{A}$ which contains $\hat{p}$. (Equivalently, $\mathcal{A}_0$ is the connected component of the Fatou set $S \smallsetminus J$ which contains $\hat{p}$. Compare 4.3.)

Corollary 8.4. Global linearization. *With $\hat{p} = f(\hat{p})$ as above, there is a holomorphic map ϕ from $\mathcal{A}$ to $\mathbb{C}$, with $\phi(\hat{p}) = 0$, so that the diagram*

$$
\begin{array}{ccc}
\mathcal{A} & \xrightarrow{\;f\;} & \mathcal{A} \\
\downarrow \phi & & \downarrow \phi \\
\mathbb{C} & \xrightarrow{\;\lambda\cdot\;} & \mathbb{C}
\end{array}
\qquad (8:3)
$$

is commutative, and so that ϕ takes a neighborhood of $\hat{p}$ bi-holomorphically onto a neighborhood of zero. Furthermore, ϕ is unique up to multiplication by a constant.

In fact, to compute $\phi(p_0)$ at an arbitrary point of $\mathcal{A}$ we must simply follow the orbit of p_0 until we reach some point p_k which is very close to $\hat{p}$, then evaluate the Kœnigs coordinate $\phi(p_k)$ and multiply by λ^{-k}. Alternatively, in terms of a local uniformizing coordinate z with $z(\hat{p}) = 0$ we can simply set $\phi(p_0) = \lim_{n\to\infty} z(f^{\circ n}(p_0))/\lambda^n$. $\quad\Box$

Now let us specialize to the case of the Riemann sphere. Suppose that $f : \hat{\mathbb{C}} \to \hat{\mathbb{C}}$ is a rational function of degree $d \geq 2$. Let $\hat{z} \in \hat{\mathbb{C}}$ be a geometrically attracting fixed point with basin of attraction $\mathcal{A} \subset \hat{\mathbb{C}}$. In some small neighborhood $\mathbb{D}_\epsilon$ of $0 \in \mathbb{C}$, note that the map $\phi : \mathcal{A} \to \mathbb{C}$ of 8.4 has a well defined holomorphic inverse $\psi_\epsilon : \mathbb{D}_\epsilon \to \mathcal{A}_0$ with $\psi_\epsilon(0) = \hat{z}$.

Lemma 8.5. Finding a critical point. *This local inverse $\psi_\epsilon : \mathbb{D}_\epsilon \to \mathcal{A}_0$ extends, by analytic continuation, to some maximal open disk $\mathbb{D}_r$ about the origin in $\mathbb{C}$. This yields a uniquely defined holomorphic map $\psi : \mathbb{D}_r \to \mathcal{A}_0$ with $\psi(0) = \hat{z}$ and $\phi(\psi(w)) \equiv w$. Furthermore, ψ extends homeomorphically over the boundary circle $\partial\mathbb{D}_r$, and the image $\psi(\partial\mathbb{D}_r) \subset \mathcal{A}_0$ necessarily contains a critical point of f.*

As an example, Figure 6 illustrates the map $f(z) = z^2 + 0.7iz$. Here the Julia set J is the outer Jordan curve, bounding the basin $\mathcal{A}$ of the attracting fixed point $\hat{z} = 0$. The critical point $c = -0.35i$ is the center of symmetry, and the fixed point $\hat{z} = 0$ is at the center of the nested circles directly above it, while the preimage $-.7i$ of the fixed point is directly below it. The curves $|\phi(z)| = \text{constant} = |\phi(c)/\lambda^n|$ have been drawn in. Thus the region $\psi(\mathbb{D}_r)$ of 8.5 is bounded by the top half of the figure eight curve through the critical point. Note that ϕ has zeros at all iterated

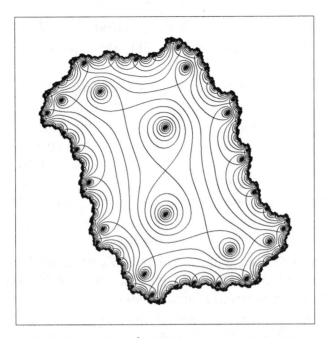

Figure 6. Julia set for $z \mapsto z^2 + .7iz$, *with curves* $|\phi| = \text{constant}$.

preimages of $\hat{z}$, and critical points (crossing points in the Figure) at all iterated preimages of the critical point c . The function $z \mapsto \phi(z)$ is unbounded, and oscillates wildly as z tends to $J = \partial \mathcal{A}$.

Proof of 8.5. Let us try to extend ψ_ϵ by analytic continuation along radial lines through the origin. It cannot be possible to extend indefinitely far in every direction, for that would yield a holomorphic map ψ from the entire complex plane onto an open set $\psi(\mathbb{C}) \subset \mathcal{A}_0 \subset \widehat{\mathbb{C}}$, with $\phi(\psi(w)) = w$. This would only be possible if the complement $\widehat{\mathbb{C}} \smallsetminus \psi(\mathbb{C})$ consisted of a single point. But the map $f|_{\psi(\mathbb{C})}$ is one-to-one, so this would imply that f is one-to-one, contradicting our hypothesis that f has degree $d \geq 2$.

Thus there must exist some largest radius r so that ψ_ϵ extends analytically throughout the open disk $\mathbb{D}_r$. Let U be the image $\psi(\mathbb{D}_r) \subset \mathcal{A}_0$. Thus we obtain a commutative diagram of conformal isomorphisms

$$
\begin{array}{ccc}
U & \xrightarrow{\ f\ } & f(U) \\
\psi \uparrow\downarrow \phi & & \psi \uparrow\downarrow \phi \\
\mathbb{D}_r & \xrightarrow{\ \lambda\cdot\ } & \lambda\mathbb{D}_r
\end{array}
$$

Note that the closure $\overline{U} \subset \widehat{\mathbb{C}}$ must be contained in the attracting basin $\mathcal{A}$. In fact, since the image of $\mathbb{D}_r$ under multiplication by λ is contained in a compact subset $\lambda\overline{\mathbb{D}}_r \subset \mathbb{D}_r$, it follows that the image $f(U)$ is contained in a corresponding compact subset $K \subset U$. It then follows by continuity that $f\left(\overline{U}\right) \subset K \subset U \subset \mathcal{A}$, which implies that $\overline{U} \subset \mathcal{A}$. In particular, it follows that ϕ is defined and holomorphic throughout a neighborhood of $\overline{U}$.

We will next show that the topological boundary ∂U contains a critical point of f. For otherwise we could analytically continue the map $\psi : \mathbb{D}_r \to \mathcal{A}$ over a strictly larger disk, as follows. For any boundary point $w_0 \in \partial\mathbb{D}_r$ choose some accumulation point z_0 in ∂U for the curve $t \mapsto \psi(tw_0)$ as $t \to 1$. If z_0 is not a critical point of f then we can choose a holomorphic branch g of f^{-1} in some neighborhood of $f(z_0)$, so that $g(f(z_0)) = z_0$, and then extend ψ holomorphically throughout a neighborhood of w_0 by the formula

$$w \mapsto g(\psi(\lambda w)) .$$

If there were no critical points at all in ∂U, then evidently these local extensions would piece together to yield a holomorphic extension of ψ through a disk which is strictly larger than $\mathbb{D}_r$.

Finally we will show that ϕ maps the compact set $\overline{U}$ homeomorphically onto the closed disk $\overline{\mathbb{D}}_r$. It suffices to show that two distinct points $z \neq z'$ in the boundary ∂U must have distinct images $\phi(z) \neq \phi(z')$ in $\partial\mathbb{D}_r$. Suppose to the contrary that $\phi(z) = \phi(z') = w \in \partial\mathbb{D}_r$. Choose a sequence of points $z_j \in U$ converging to z and a sequence of points $z'_j \in U$ converging to z'. Then the sequences $\{\phi(z_j)\}$ and $\{\phi(z'_j)\}$ converge to the same limit in $\partial\mathbb{D}_r$. Let L_j be the straight line segment from $\phi(z_j)$ to $\phi(z'_j)$ in $\mathbb{D}_r$, and let $X \subset \partial U$ be the set of all accumulation points for the curves $\psi(L_j)$ as $j \to \infty$. Then it is not difficult to show that X is a compact connected set containing both z and z', but that $f(X)$ consists of a single point in U. Evidently this is impossible. $\square$

More generally, if $\mathcal{O} = \{z_1, \ldots, z_m\}$ is an attracting periodic orbit of period m, so that each z_j is an attracting fixed point for the m-fold composition $f^{\circ m}$, then the immediate basin $\mathcal{A}_0 = \mathcal{A}_0(\mathcal{O}, f)$ is defined to be the union of the immediate attractive basins $\mathcal{A}_0(z_j)$ of the m fixed points $z_j = f^{\circ m}(z_j)$ under $f^{\circ m}$.

The following fundamental result is due to Fatou and Julia.

Theorem 8.6. Finding periodic attractors. *If f is a rational map of degree $d \geq 2$, then the immediate basin of every attracting periodic orbit contains at least one critical point. Hence the number of attracting periodic orbits is finite, less than or equal to the number of critical points.*

Proof. In the case of a geometrically attracting fixed point, the first statement follows immediately from 8.5; while a superattracting fixed point is itself the required critical point in its basin. Now consider a period m attracting orbit $\{z_j\}$ with $f(z_j) = z_{j+1}$, taking the subscripts j to be integers modulo m. Evidently $f(\mathcal{A}_0(z_j)) \subset \mathcal{A}_0(z_{j+1})$. If none of the $\mathcal{A}_0(z_j)$ contained a critical point, then, by the chain rule, the m-fold composition mapping each $\mathcal{A}_0(z_j)$ into itself would not have any critical point, which is impossible.

The conclusion now follows since the attractive basins of the various periodic attractors are clearly pairwise disjoint, and since a non-constant rational map can have only finitely many critical points. $\square$

(For another proof, see Problem 8-g.)

As an example, for a polynomial map of degree $d \geq 2$, there are at most $d - 1$ finite critical points, and hence at most $d - 1$ periodic attractors (not counting the fixed point at infinity). In the case of a rational map, there are $2d - 2$ critical points, counted with multiplicity. This follows from the Riemann-Hurwitz formula 5.1, or by simply inspecting the required polynomial equation $p'q - q'p = 0$ where $f(z) = p(z)/q(z)$, taking particular care with the possibility of a critical point at infinity. Hence if $d \geq 2$ there are at most $2d - 2$ periodic attractors. (Compare 10.12, 13.2.)

By way of contrast, for two complex dimensions, Newhouse has shown that a polynomial automorphism of $\mathbb{C}^2$ (or of $\mathbb{R}^2$) can actually have infinitely many periodic attractors. In this case, there are no critical points to work with.

Note that this theorem gives a constructive algorithm for locating the attracting periodic points, if they exist, for any non-linear rational map. Starting at each one of the critical points, simply iterate the map many times and then test for (approximate) periodicity. (Of course if the period is very large, then this becomes impractical. As an explicit example, it is easy to check that the quadratic map $f(z) = z^2 - 1.5$ of Figure 11 has no attracting orbits of reasonable period. However, I know no way of deciding whether it has an attracting orbit of some very high period.)

Theorem 8.7. Topology of $\mathcal{A}_0$. *Let $\mathcal{A}_0$ be the immediate basin of an attracting fixed point (either geometrically attracting or superattracting). Then the complement $\widehat{\mathbb{C}} \smallsetminus \mathcal{A}_0$ is either connected or else has uncountably many connected components.*

(Compare 4.12.) It follows that $\mathcal{A}_0$ itself is either simply connected or infinitely connected. For an example of the infinitely connected case, see Figure 1b. Note that the analogous statement for an attracting periodic point of period p follows by applying 8.7 to the iterate $f^{\circ p}$.

Proof of 8.7. Choose a small open disk N_0 about the attracting point $\hat{z}$ so that $f(\overline{N}_0) \subset N_0$, and so that the boundary ∂N_0 is a simple closed curve containing no iterated forward images of critical points. Let N_k be the connected component of $f^{-k}(N_0)$ which contains N_0. Thus $N_0 \subset N_1 \subset N_2 \subset \cdots$ with union $\mathcal{A}_0$. In fact, any point of $\mathcal{A}_0$ can be joined to $\hat{z}$ by a path $P \subset \mathcal{A}_0$. Choosing k so that $f^{\circ k}(P) \subset N_0$, we see inductively that $f^{\circ(k-i)}(P) \subset N_i$, and hence that $P \subset N_k$. Evidently each N_k is bounded by some finite number of simple closed curves.

Case 1. If each N_k is bounded by one simple closed curve, then $\widehat{\mathbb{C}} \smallsetminus N_k$ is connected, and $\widehat{\mathbb{C}} \smallsetminus \mathcal{A}_0$, being the intersection of a nested sequence of connected sets, is itself connected.

Case 2. Suppose that some N_k is bounded by two or more simple closed curves $\Gamma_1, \ldots, \Gamma_m$. Then each connected component of $N_{2k} \smallsetminus \overline{N}_k$ is a branched covering of $N_k \smallsetminus \overline{N}_0$. Since $N_k \smallsetminus \overline{N}_0$ has $m+1$ boundary curves, it follows that each such component has at least $m+1$ boundary curves, of which exactly one must coincide with one of the Γ_i. It follows that each of the m connected components of $\widehat{\mathbb{C}} \smallsetminus N_k$ contains at least m connected components of $\widehat{\mathbb{C}} \smallsetminus N_{2k}$. Similarly, each of these contains at least m connected components of N_{3k}, and so on. Proceeding as in the proof of 4.12, it follows that $\widehat{\mathbb{C}} \smallsetminus \mathcal{A}_0$ has uncountably many components. $\square$

Repelling Points.

For most purposes we can simply define a "repelling" fixed point to be one with multiplier satisfying $|\lambda| > 1$. However, it is more satisfying to have a topologically invariant characterization.

Definition. A fixed point $\hat{p} = f(\hat{p})$ of a continuous map will be called *topologically repelling* if there is a neighborhood U of $\hat{p}$ so that for every $p \neq \hat{p}$ in U there exists some $n \geq 1$ so that the n-th forward image $f^{\circ n}(p)$ lies outside of U. In other words, the only infinite orbit $p_0 \mapsto p_1 \mapsto p_2 \mapsto \cdots$ which is completely contained in U must be the orbit of the fixed point itself. Such a U is called a *forward isolating* neighborhood of $\hat{p}$.

Lemma 8.8. Characterization of topologically repelling points. *A fixed point of a holomorphic map is topologically repelling if and only if its multiplier satisfies $|\lambda| > 1$.*

Proof. If $|\lambda| > 1$, then it follows from 8.2 (or from a much more elementary exercise in calculus) that the point is topologically repelling. I am indebted to S. Zakeri for the following proof of the converse statement. If $\hat{p}$ is a topologically repelling point for f, note first that $\lambda \neq 0$ (and in fact $|\lambda| \geq 1$), since $\hat{p}$ clearly cannot be both attracting and repelling. Thus we can choose a compact forward isolating neighborhood N, which is small enough so that f maps N homeomorphically onto some compact neighborhood $f(N)$ of $\hat{p}$. Let

$$N_k \;=\; N \cap f^{-1}(N) \cap \cdots \cap f^{-k}(N)$$

be the compact neighborhood consisting of points for which the first k forward images all belong to N. Thus $N = N_0 \supset N_1 \supset N_2 \supset \cdots$, with intersection the single point $\hat{p}$ since N is an isolating neighborhood. By compactness, it follows that the diameter of N_k tends to zero as $k \to \infty$. But it follows immediately from the construction that

$$f(N_k) \;=\; N_{k-1} \cap f(N) \,,$$

where $N_{k-1} \subset f(N)$ for k large since the diameters tend to zero. Thus $f(N_k) = N_{k-1}$ for k large; in fact f maps N_k homeomorphically onto N_{k-1}. Now let U_k be the connected component of the interior of N_k which contains $\hat{p}$. Then it follows that f^{-1} maps U_{k-1} biholomorphically onto the strictly smaller set U_k. By the Schwarz Lemma, its multiplier must satisfy $|\lambda^{-1}| < 1$, which proves that $|\lambda| > 1$ as required. $\square$

Remark. Lemmas 8.1 and 8.8 work only over the complex numbers. Over the real numbers, examples such as $f(x) = x \pm x^3$ show that a fixed point with multiplier $\lambda = 1$ may perfectly well be topologically attracting or topologically repelling.

Figure 7. Detail of Julia set for $z \mapsto z^2 - .744336 + .121198i$.

The Kœnigs Linearization Theorem 8.2, in the repelling case, helps us to understand why the Julia set $J(f)$ is so often a complicated "fractal" set.

Corollary 8.9. *Suppose that the rational function f has a repelling periodic point $\hat{z}$ for which the multiplier λ is not a real number. Then $J(f)$ cannot be a smooth manifold, unless it is all of $\widehat{\mathbb{C}}$.*

To see this, choose any point $z_0 \in J(f)$ which is close to $\hat{z}$, and let $w_0 = \phi(z_0)$. Then $J(f)$ must also contain an infinite sequence of points $z_0 \leftarrow z_1 \leftarrow z_2 \leftarrow \dots$ with Kœnigs coordinates $\phi(z_n) = w_0/\lambda^n$ which lie along a logarithmic spiral and converge to zero. Evidently such a set can not lie in any smooth one-dimensional submanifold of $\mathbb{C}$. $\square$

In fact, if we recall that the iterated pre-images of our periodic point are everywhere dense in $J(f)$, then we see that such sequences lying on logarithmic spirals are extremely pervasive. Compare Figures 7 and 8 which show typical examples of such spiral structures, associated with repelling points of periods 2 and 1 respectively.

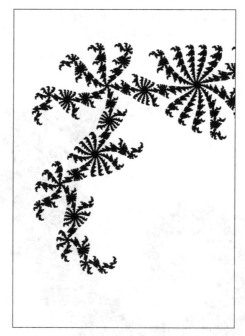

Figure 8. Detail of Julia set for $z \mapsto z^2 + .424513 + .207530i$.

The global form of the Linearization Theorem in the repelling case is rather different from the statement 8.4 in the geometrically attracting case. In particular, there is no analogous concept of a "repelling basin", and no analogous extension $S \supset A \to \mathbb{C}$. However, we can extend ϕ^{-1} to a map $\mathbb{C} \to S$.

Corollary 8.10. Global extension of ϕ^{-1}. *If $\hat{p}$ is a repelling fixed point for the holomorphic map $f : S \to S$, then there is a holomorphic map $\psi : \mathbb{C} \to S$, with $\psi(0) = \hat{p}$, so that the diagram*

$$
\begin{array}{ccc}
S & \xrightarrow{\ f\ } & S \\
\uparrow \psi & & \uparrow \psi \\
\mathbb{C} & \xrightarrow{\ \lambda\cdot\ } & \mathbb{C}
\end{array}
$$

is commutative, and so that ψ maps a neighborhood of zero biholomorphically onto a neighborhood of $\hat{p}$. Here ψ is unique except that it may be replaced by $w \mapsto \psi(cw)$ for any constant $c \neq 0$.

For ϵ sufficiently small, let $\psi_\epsilon : \mathbb{D}_\epsilon \to S$ be that branch of ϕ^{-1} which maps zero to $\hat{p}$. Now to compute $\psi(w)$ for any $w \in \mathbb{C}$, we simply choose n large enough so that $w/\lambda^n \in \mathbb{D}_\epsilon$, and then set $\psi(w) = f^{\circ n}(\psi_\epsilon(w/\lambda^n))$. Details will be left to the reader. □

Concluding Problems.

Problem 8-a. The identification torus of a fixed point. Suppose that f has a geometrically attracting or repelling fixed point $\hat{p}$ with multiplier λ. Let U be any neighborhood of $\hat{p}$ small enough so that f maps U biholomorphically, with $f(U) \subset U$ in the attracting case or $f(U) \supset U$ in the repelling case. Form an identification space $T = (U \smallsetminus \{\hat{p}\})/f$ by identifying p with $f(p)$ whenever both belong to U. Show that T is a Riemann surface, independent of the choice of U, with the topology of a torus. Show in fact that T is conformally isomorphic to the quotient $\mathbb{C}/\Lambda$, where Λ is the lattice $2\pi i\mathbb{Z} \oplus (\log \lambda)\mathbb{Z}$.

Problem 8-b. Global linearization. Suppose that $f : S \to S$ is a holomorphic map from the Riemann surface S onto itself. (For example suppose that $S = \hat{\mathbb{C}}$, so that f is a rational map.) Show that the linearizing map ϕ of 8.4 maps the attracting basin A onto $\mathbb{C}$. Show that $p_0 \in A$ is a critical point of ϕ if and only if the orbit $f : p_0 \mapsto p_1 \mapsto p_2 \mapsto \cdots$ contains some critical point of f.

Problem 8-c. Asymptotic values. In order to extend Theorem 8.6 to a non-compact Riemann surface such as $\mathbb{C}$ or $\mathbb{C} \smallsetminus \{0\}$, we need some definitions. Let $f : S \to S'$ be a holomorphic map between Riemann surfaces. A point $v \in S'$ is a *critical value* if it is the image under f of a *critical point*, that is a point at which the first derivative of f vanishes. It is an *asymptotic value* if there exists a continuous path $[0,1) \to S$ which "diverges to infinity" in S, or in other words eventually leaves any compact subset of S, but whose image under f converges to the point v. Recall from §2 that a connected open set $U \subset S'$ is *evenly covered* if every component of $f^{-1}(U)$ maps homeomorphically onto U, and that f is a *covering map* if every point of S' has a neighborhood which is evenly covered.

Show that a simply connected open subset of S' is evenly covered by f if and only if it contains no critical value or asymptotic value. (Compare [Goldberg and Keen].) In particular, f is a covering map if and only if S' contains no critical values and no asymptotic values.

For a holomorphic self-map $f : S \to S$, show that the immediate basin of any attracting periodic orbit must contain either a critical value or an asymptotic value or both, except in the special case of a linear map from $\mathbb{C}$ or $\widehat{\mathbb{C}}$ to itself. As an example, for any $c \neq 0$, show that the transcendental map $f(z) = ce^z$ from $\mathbb{C}$ to itself has no critical points, and just one asymptotic value, namely $z = 0$. Conclude that it has at most one periodic attractor. If $|c| < 1/e$ show that f maps the unit disk into itself, and that f has an attracting fixed point in this disk.

The map f is *proper*, if the pre-image $f^{-1}(K)$ of every compact set $K \subset S'$ is a compact subset of S. (If S is compact, then every map on S is proper.) Show that a proper map has no asymptotic values.

Problem 8-d. Topological attraction and repulsion. Suppose that X is a locally compact space and that f maps a compact neighborhood N of $\hat{x}$ homeomorphically onto a compact neighborhood N', with $f(\hat{x}) = \hat{x}$. Show that f is topologically repelling at $\hat{x}$ if and only if f^{-1} is topologically attracting at $\hat{x}$. (Here the hypothesis that f is locally one-to-one is essential. For example the map $f(z) = z^2$ is attracting at the origin, and the non-smooth map $g(z) = 2z^2/|z|$ is repelling at the origin, although neither one has a local inverse.)

Problem 8-e. The image $\psi(\mathbb{C}) \subset S$. If $\hat{p}$ is a repelling point for the holomorphic map $f : S \to S$, show that the image of the map $\psi : \mathbb{C} \to S$ of 8.10 is everywhere dense, and in fact that the complement $S \smallsetminus \psi(\mathbb{C})$ consists of grand orbit finite points. (Compare 4.7. There are at most two such points when $S = \widehat{\mathbb{C}}$, at most one when $S = \mathbb{C}$, and none for other non-hyperbolic surfaces.)

Problem 8-f. Counting basin components. Let $\mathcal{A}$ be the attracting basin of a periodic point which may be either superattracting or geometrically attracting. If some connected component of $\mathcal{A}$ is not periodic, show that $\mathcal{A}$ has infinitely many components. Suppose then that $\mathcal{A}$ has only finitely many components forming a periodic cycle. If these components are simply connected, use the Riemann-Hurwitz formula 7.2 to show that the period is at most two. (Example: $f(z) = 1/z^2$.) If they are infinitely connected, show that the period must be one.

Problem 8-g. Critical points in the basin. Give another proof of 8.6 as follows. Suppose there were an attracting periodic orbit $\mathcal{O}$ with no critical point in its immediate basin. If U is a small neighborhood of a point $p \in \mathcal{O}$, show that for each $k \geq 1$ there would be a unique branch $g_k : U \to \hat{\mathbb{C}}$ of $f^{-k}|_U$ which maps p into $\mathcal{O}$. Show that the family $\{g_k\}$ would have to be normal; which is impossible since the first derivative of g_k at p must be unbounded.

§9. Böttcher's Theorem and Polynomial Dynamics

This section studies the *superattracting* case, with multiplier λ equal to zero. As usual, we can choose a local uniformizing parameter z with fixed point $z = 0$. Thus our map takes the form

$$f(z) = a_n z^n + a_{n+1} z^{n+1} + \cdots , \qquad (9:1)$$

with $n \geq 2$ and $a_n \neq 0$.

9.1. Theorem of Böttcher.[*] *With f as above, there exists a local holomorphic change of coordinate $w = \phi(z)$, with $\phi(0) = 0$, which conjugates f to the n-th power map $w \mapsto w^n$ throughout some neighborhood of zero. Furthermore, ϕ is unique up to multiplication by an $(n-1)$-st root of unity.*

Thus near any critical fixed point, f is conjugate to a map of the form

$$\phi \circ f \circ \phi^{-1} : w \mapsto w^n ,$$

with $n \geq 2$. This Theorem is often applied in the case of a fixed point at infinity. For example, any polynomial of degree $n \geq 2$ has a superattracting fixed point at infinity. (Compare 9.5 below.)

Proof of existence. The proof will be quite similar to the proof of 8.2. With f as in $(9:1)$, let us first choose a solution c to the equation $c^{n-1} = a_n$. Then the linearly conjugate map $cf(z/c)$ will have leading coefficient equal to $+1$. Thus we may assume, without loss of generality, that our map has the form $f(z) = z^n(1 + b_1 z + b_2 z^2 + b_3 z^3 + \cdots)$, or briefly

$$f(z) = z^n (1 + \eta(z)) \qquad \text{with} \qquad \eta(z) = b_1 z + b_2 z^2 + \cdots . \qquad (9:2)$$

Choose a radius $0 < r < 1/2$ which is small enough so that $|\eta(z)| < 1/2$ on the disk $\mathbb{D}_r$ of radius r. Then clearly f maps this disk into itself, with $|f(z)| \leq \frac{3}{4}|z|$ and with $f(z) \neq 0$ for $z \in \mathbb{D}_r \smallsetminus \{0\}$. The k-fold iterate $f^{\circ k}$ also maps $\mathbb{D}_r$ into itself, and we see inductively that it has the form $f^{\circ k}(z) = z^{n^k}(1 + n^{k-1}b_1 z + (\text{higher terms}))$. The idea of the proof is to set

$$\phi_k(z) = \sqrt[n^k]{f^{\circ k}(z)} = z(1 + n^{k-1}b_1 z + \cdots)^{1/n^k} = z(1 + \frac{b_1}{n} z + \cdots) ,$$

---

[*] Proved in 1904. L. E. Böttcher was born in Warsaw in 1878. He took his doctorate in Leipzig in 1898, working in Iteration Theory, and then moved to Lvov, where he retired in 1935. He published in Polish and Russian. (The Russian form of his name is Бётхер.)

choosing that n^k-th root which has derivative $+1$ at the origin. Clearly $\phi_k(f(z)) = \phi_{k+1}(z)^n$. We will show that the functions ϕ_k converge uniformly to a limit function $\phi : \mathbb{D}_r \to \mathbb{D}$ which satisfies the required equation $\phi(f(z)) = \phi(z)^n$. In order to prove convergence, let us make the substitution $z = e^Z$, where Z ranges over the left half-plane $\mathrm{Re}(Z) < \log r$. Then the map f in the disk $\mathbb{D}_r$ corresponds to a map $F(Z) = \log f(e^Z)$ in this left half-plane. This can be written more precisely as

$$F(Z) = \log\left(e^{nZ}(1+\eta)\right) = nZ + \log(1+\eta)$$
$$= nZ + (\eta - \eta^2/2 + \eta^3/3 - + \cdots),$$

where $\eta = \eta(e^Z) \in \mathbb{D}_{1/2}$. (In this form, we have made an explicit choice as to which branch of the logarithm to use.) Evidently F is a well defined holomorphic function, which maps the half-plane $\mathrm{Re}(Z) < \log r$ into itself. Since $|\eta| < 1/2$, we have

$$|F(Z) - nZ| = |\log(1+\eta)| < \log 2 < 1 \qquad (9:3)$$

for all Z in this half-plane.

Similarly, the map $\phi_k(z) = f^{\circ k}(z)^{1/n^k}$ in the z-plane corresponds to a map

$$\Phi_k(Z) = \log \phi_k(e^Z) = F^{\circ k}(Z)/n^k$$

which is defined and holomorphic throughout the half-plane $\mathrm{Re}(Z) < \log r$. By $(9:3)$, we have

$$|\Phi_{k+1}(Z) - \Phi_k(Z)| = |F^{\circ k+1}(Z) - n\,F^{\circ k}(Z)|/n^{k+1} < 1/n^{k+1}.$$

(It follows that the Φ_k converge uniformly to a limit function Φ, with $\Phi(F(z)) = n\Phi(Z)$.) Since the exponential map from the left half-plane onto $\mathbb{D}$ reduces distances, it follows that

$$|\phi_{k+1}(z) - \phi_k(z)| < 1/n^{k+1}$$

for $|z| < r$. Therefore, as $k \to \infty$ the sequence of holomorphic functions $z \mapsto \phi_k(z)$ on the disk $|z| < r$ converges uniformly to a holomorphic limit $\phi(z)$. Clearly ϕ satisfies the required identity $\phi(f(z)) = \phi(z)^n$.

Proof of Uniqueness. It suffices to study the special case $f(z) = z^n$. If a map of the form $\phi(z) = c_1 z + c_k z^k + \text{(higher terms)}$ conjugates $z \mapsto z^n$ to itself, then the series

$$\phi(z^n) = c_1 z^n + c_k z^{nk} + \cdots$$

must be equal to

$$\phi(z)^n = c_1^n z^n + n c_1^{n-1} c_k z^{n+k-1} + \cdots,$$

with $nk > n + k - 1$. Comparing coefficients, we find that $c_1^{n-1} = 1$ and that all higher coefficients are zero. □

Remark. Given a global holomorphic map $f : S \to S$ with a superattracting point $\hat{p}$, we can choose a local uniformizing parameter $z = z(p)$ with $z(\hat{p}) = 0$, and construct the Böttcher coordinate $w = \phi(z(p))$ as above. *To simplify the notation, we will henceforth forget the intermediate parameter z and simply write $w = \phi(p)$.*

In analogy with 8.4, one might hope that the local mapping $p \mapsto \phi(p)$ could be extended throughout the entire basin of attraction of $\hat{p}$ as a holomorphic mapping $\mathcal{A} \to D$. However, this is not always possible. Such an extension would involve computing expressions of the form

$$p \mapsto \sqrt[n]{\phi(f^{\circ n}(p))} \,,$$

and this may not work, since the n-th root cannot always be defined as a single valued function. For example, there is trouble whenever some other point in the basin maps exactly onto the superattractive point, or whenever the basin is not simply-connected. However, if we consider only the absolute value of ϕ, then there is no problem.

9.2. Corollary. Extension of $|\phi|$. *If $f : S \to S$ has a superattracting fixed point $\hat{p}$ with basin $\mathcal{A}$, then the function $p \mapsto |\phi(p)|$ of 9.1 extends uniquely to a continuous map $|\phi| : \mathcal{A} \to [0, 1)$ which satisfies the identity $|\phi(f(p))| = |\phi(p)|^n$.*

Proof. Set $|\phi(p)|$ equal to $|\phi(f^{\circ k}(p))|^{1/n^k}$ for large k. □

Just as in 8.5, we can start with a local inverse ψ_ϵ to ϕ, which is defined on a disk of radius ϵ and satisfies $\psi_\epsilon(0) = \hat{p}$, and then extend as far as possible by analytic continuation. In this way we prove the following.

9.3. Theorem. Critical points in the basin. *Let $f : \hat{\mathbb{C}} \to \hat{\mathbb{C}}$ be a rational function with superattracting fixed point $\hat{p}$, and let $\mathcal{A}_0$ be the immediate attracting basin of $\hat{p}$. Then there are two possibilities:*

Case 1. *The Böttcher map extends to a conformal isomorphism from $\mathcal{A}_0$ onto the open unit disk $\mathbb{D}$, which necessarily conjugates $f|_{\mathcal{A}_0}$ to the n-th power map $w \mapsto w^n$ on $\mathbb{D}$. In this case f evidently has no critical points other than $\hat{p}$ in $\mathcal{A}_0$.*

Case 2. *Otherwise, there exists a maximal number $0 < r < 1$ so that the local inverse $\psi_\epsilon : \mathbb{D}_\epsilon \to \mathcal{A}_0$ extends to a conformal isomorphism ψ from the open disk $\mathbb{D}_r$ of radius r onto an open subset $U = \psi(\mathbb{D}_r) \subset \mathcal{A}_0$. In this case, the closure $\overline{U}$ is a compact subset of $\mathcal{A}_0$, and the boundary $\partial U \subset \mathcal{A}_0$ contains* *at least one critical point of f .*

Compare Figure 9 which illustrates Case 1 and Figure 10 which illustrates Case 2.

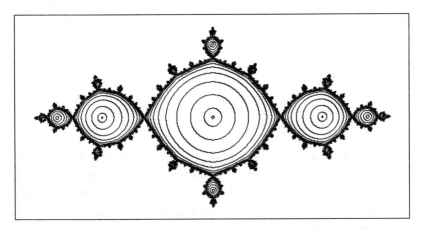

Figure 9. Julia set for $f(z) = z^2 - 1$. The 4-th degree map $f \circ f$ has two superattracting fixed points at $z = 0$ and $z = -1$, with no other critical points in the immediate basins. The grand orbit of a representative curve $|\phi| = $ constant has been drawn in for both attracting basins. Note that each such curve in an immediate basin $\mathcal{A}_0$ maps to the next smaller curve in $\mathcal{A}_0$ by a 2-fold covering.

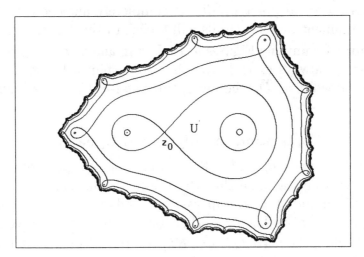

Figure 10. Julia set for the map $f(z) = z^3 + z^2$, which has a critical point $z_0 = -2/3$ in the immediate basin of the superattracting point $z = 0$. The grand orbit of the curve $|\phi| = $ constant through z_0 has been drawn in.

Proof of 9.3. As in the proof of 8.5, we can extend ψ_ϵ by analytic continuation, either to the unit disk $\mathbb{D}$ of radius $r = 1$, or to some maximal disk $\mathbb{D}_r$ of radius $r < 1$. First note that the resulting map $\psi = \psi_r :$ $\mathbb{D}_r \to U \subset \mathcal{A}_0$ has no critical points. For if there were a point $w \in \mathbb{D}_r$ with $\psi'(w) = 0$ then certainly $w \neq 0$. Hence the equation

$$\psi(w^n) = f(\psi(w)) \qquad (9:4)$$

would imply that w^n is also a critical point. This would yield a sequence of critical points $w, w^n, w^{n^2}, \ldots$ tending to zero, which is impossible. Thus ψ is locally one-to-one; and the set of all pairs $w_1 \neq w_2$ with $\psi(w_1) = \psi(w_2)$ forms a closed subset of $\mathbb{D}_r \times \mathbb{D}_r$.

We must show that ψ is actually one-to-one on $\mathbb{D}_r$. If $\psi(w_1) = \psi(w_2)$, then it follows from 9.2 that $|w_1| = |\phi(\psi(w_1))|$ is equal to $|w_2|$. If there were such a pair with $w_1 \neq w_2$, then we could choose it with $|w_1| = |w_2|$ minimal. But ψ is an open mapping, so for any w_1' sufficiently close to w_1 we could choose w_2' close to w_2 with $\psi(w_1') = \psi(w_2')$. In particular, we could choose the w_j' with $|w_j'| < |w_j|$, which contradicts the choice of $|w_j|$.

Suppose that $r = 1$. Then the function $p \mapsto |\phi(p)|$ of 9.2 clearly tends to the limit $+1$ as p tends to the boundary of U. yet takes values strictly less than 1 throughout $\mathcal{A}$. It follows that every boundary point of U lies outside of $\mathcal{A}$, which implies that $U = \mathcal{A}_0$.

Now suppose that $r < 1$. Then the proof that $\partial U \subset \mathcal{A}_0$, and that there exists a critical point in ∂U is completely analogous to the corresponding argument in 8.5. Details will be left to the reader. $\square$

Caution: Examining Figures 9, 10, and in analogy with 8.5, one might expect that ϕ always extends to a homeomorphism between the closure $\overline{U}$ and the closed disk $\overline{D}_r$; however, this is false. Compare Figures 11, 12 below.

Application to Polynomial Dynamics.

Let $f(z) = a_n z^n + a_{n-1} z^{n-1} + \cdots + a_1 z + a_0$ be a polynomial of degree $n \geq 2$. (We always assume that the leading coefficient a_n is non-zero.) Then we will see that f has a superattracting fixed point at infinity, so that we can apply Böttcher's Theorem. But first a more elementary construction.

By definition, the *filled Julia set* $K(f)$ is the set of all $z \in \mathbb{C}$ for which the orbit of z under f is bounded.

9.4. Lemma. The filled Julia set. *For any polynomial f of degree at least two, this set $K = K(f) \subset \mathbb{C}$ is compact, with connected complement. It can be described as the union of the*

Julia set $J = J(f)$ together with all bounded components of the complement $\mathbb{C} \smallsetminus J$. On the other hand, J can be described as the boundary ∂K.

(A compact subset of $\mathbb{C}$ is called *full* if its complement is connected. Any compact subset of $\mathbb{C}$ can be "filled" by adjoining all bounded components of its complement.)

Proof of 9.4. It suffices to consider the special case of a *monic* polynomial, that is one with leading coefficient a_n equal to $+1$. For if $g(z)$ is a polynomial with leading coefficient $a_n \neq 1$, then choosing some solution c to the equation $c^{n-1} = a_n$, we see that the linearly conjugate polynomial $f(z) = cg(z/c)$ is monic. It is then not difficult to find a constant b so that $|f(z)| > |z|$ whenever $|z| > b$. For example, assuming that f is monic, we can take $b = 3 \max_k |a_k| \geq 3$. If $|z| > b$, it follows that

$$|f(z)/z^n| = \left| 1 + \frac{a_{n-1}}{z} + \cdots + \frac{a_0}{z^n} \right| \geq 1 - \frac{1}{3} - \frac{1}{9} - \cdots = \frac{1}{2},$$

so that $|f(z)| \geq |z^n|/2 \geq \frac{3}{2}|z|$. It follows that every z with $|z| > b$ belongs to the attracting basin $\mathcal{A} = \mathcal{A}(\infty)$ of the point at infinity. Evidently K can be identified with the complement $\widehat{\mathbb{C}} \smallsetminus \mathcal{A}$. Hence K is compact, and $\partial K = \partial \mathcal{A}$ is equal to the Julia set by 4.9.

We must show that $\mathcal{A}$ is connected. If U is any bounded component of the $\mathbb{C} \smallsetminus J$, we will show that $|f^{\circ n}(z)| \leq b$ for every $z \in U$ and every $n \geq 0$. For otherwise, by the Maximum Modulus Principle, there would be some $z \in \partial U \subset J$ with $|f^{\circ n}(z)| > b$. But this would imply that $z \in \mathcal{A}$, which is impossible. Thus every bounded component of $\mathbb{C} \smallsetminus J$ is contained in the filled Julia set K, and the unique unbounded component can be identified with $\mathbb{C} \smallsetminus K = \mathbb{C} \cap \mathcal{A}(\infty)$. $\square$

To better understand this filled Julia set K, we consider the dichotomy of Theorem 9.3 for the complementary domain $\mathcal{A}(\infty) = \widehat{\mathbb{C}} \smallsetminus K$. This yields the following.

9.5. Theorem. Connected K $\Leftrightarrow$ bounded critical orbits. *Let f be a polynomial of degree $d \geq 2$. If the filled Julia set $K = K(f)$ contains all of the finite critical points of f, then both K and $J = \partial K$ are connected, and the complement of K is conformally isomorphic to the exterior of the closed unit disk $\overline{\mathbb{D}}$ under an isomorphism*

$$\Phi : \mathbb{C} \smallsetminus K \longrightarrow \mathbb{C} \smallsetminus \overline{\mathbb{D}}$$

which conjugates f on $\mathbb{C} \smallsetminus K$ to the n-th power map $w \mapsto w^n$. On the other hand, if at least one critical point of f belongs to

$\mathbb{C} \setminus K$, then both K and J have uncountably many connected components.

Compare Figure 11 which illustrates the first possibility and Figure 12 which illustrates the second. The proof of 9.5 will be based on the following. To study the behavior of f near infinity, we make the usual substitution $\zeta = 1/z$ and consider the rational function

$$F(\zeta) = \frac{1}{f(1/\zeta)} .$$

Again we may assume that f is monic. From the asymptotic equality $f(z) \sim z^n$ as $z \to \infty$, it follows that $F(\zeta) \sim \zeta^n$ as $\zeta \to 0$. Thus F has a superattracting fixed point at $\zeta = 0$. (More explicitly, it is not difficult to derive a power series expansion of the form

$$F(\zeta) = \zeta^n - a_{n-1}\zeta^{n+1} + (a_{n-1}^2 - a_{n-2})\zeta^{n+2} + \cdots$$

for $|\zeta|$ small.) There is an associated Böttcher function

$$\phi(\zeta) = \lim_{k \to \infty} F^{\circ k}(\zeta)^{1/n^k} \in \mathbb{D} ,$$

which is defined and biholomorphic for $|\zeta|$ sufficiently small, with $\phi'(0) = 1$ since f is assumed to be monic. In practice, it is more convenient to work with the reciprocal

$$\Phi(z) = \frac{1}{\phi(1/z)} = \lim_{k \to \infty} f^{\circ k}(z)^{1/n^k} \in \mathbb{C} \setminus \overline{\mathbb{D}} .$$

Thus Φ maps some neighborhood of infinity biholomorphically onto a neighborhood of infinity, with $\Phi(z) \sim z$ as $|z| \to \infty$, and Φ conjugates the degree n polynomial map f to the n-th power map, so that

$$\Phi(f(z)) = \Phi(z)^n . \tag{9 : 5}$$

Proof of 9.5. (Compare Problem 9-c.) Suppose first that there are no critical points other than ∞ in the attracting basin $A = A(\infty)$. Then by 9.3 the Böttcher map extends to a conformal isomorphism $A \xrightarrow{\cong} \mathbb{D}$. It follows that the function Φ extends to a conformal isomorphism $\mathbb{C} \setminus K \xrightarrow{\cong} \mathbb{C} \setminus \overline{\mathbb{D}}$. Now each annulus

$$\mathbb{A}_{1+\epsilon} = \{z \in \mathbb{C} ; 1 < |z| < 1 + \epsilon\}$$

maps under $\Psi = \Phi^{-1}$ to a connected set $\Psi(\mathbb{A}_{1+\epsilon}) \subset \mathbb{C} \setminus K$. The closure $\overline{\Psi(\mathbb{A}_{1+\epsilon})}$ is a compact connected set which evidently contains the Julia set

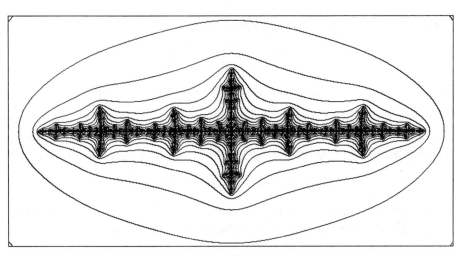

Figure 11. Julia set for $f(z) = z^2 - 3/2$. *An equipotential*

$$G = \log|\phi| = \text{constant}$$

has been drawn in, together with its iterated forward and backward images. (Compare 9.6.) Each such curve maps to the next larger curve by a two-fold covering.

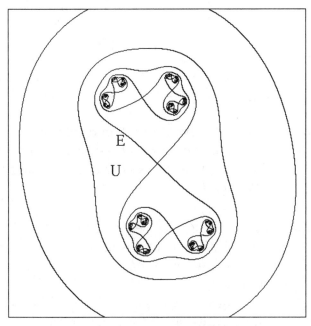

Figure 12. The Julia set for $f(z) = z^2 + (1 + i/2)$ *is totally disconnected (a Cantor set). The neighborhood of infinity* $U = \Psi(\mathbb{C} \smallsetminus \overline{\mathbb{D}}_r)$ *is the complement of the region bounded by the figure eight equipotential curve* E *through the critical point* $z = 0$. *Note that the two components of* $f^{-1}(E)$ *are also figure eight curves, as are the four components of* $f^{-2}(E)$, *and so on. However, the iterated forward images of* E *are all topological circles.*

$J = \partial \mathcal{A}$. It follows that the intersection

$$J \;=\; \bigcap_{\epsilon>0} \overline{\Psi(\mathbb{A}_{1+\epsilon})}$$

is also connected, and it then follows easily from 9.4 that K is connected.

Now suppose that there is at least one critical point in $\mathbb{C} \smallsetminus K$. Then the conclusion of 9.3 translates as follows: *There is a smallest number $r > 1$ so that the inverse of Φ near infinity extends to a conformal isomorphism*

$$\Psi \;:\; \mathbb{C} \smallsetminus \overline{\mathbb{D}}_r \;\xrightarrow{\;\cong\;}\; U \subset \mathbb{C} \smallsetminus K \;.$$

Furthermore the boundary ∂U of this open set $U = \Psi(\mathbb{C} \smallsetminus \overline{\mathbb{D}}_r)$ is a compact subset of $\mathbb{C} \smallsetminus K$ which contains at least one critical point of f.

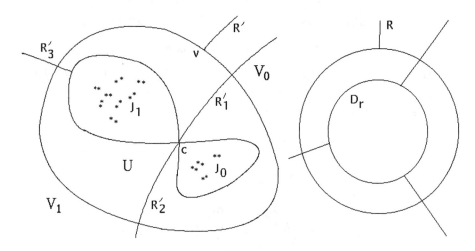

Figure 13. Sketch for the proof of 9.5 in degree $n = 3$, with the z-plane on the left and the $w = \Phi(z)$-plane on the right. The open set $U = \Psi(\mathbb{C} \smallsetminus \overline{\mathbb{D}}_r)$ is the exterior of the region bounded by the figure eight through the critical point.

We will show that the closure $\overline{U}$ separates the plane into two or more bounded open sets, each of which contains uncountably many points of the Julia set. Let c be a critical point in ∂U. Then the corresponding *critical value* $v = f(c)$ clearly belongs to U, with $|\Phi(v)| = r^n > r$. Consider the infinite *ray* $R \subset \mathbb{C} \smallsetminus \mathbb{D}_r$ consisting of all products $t\Phi(v)$ with $t \geq 1$. The image $R' = \Psi(R) \subset U$ is called an *external ray* to the point v, associated with the compact set $K \subset \mathbb{C}$.

Now consider the full inverse image $f^{-1}(R') \subset \overline{U}$. Clearly the intersection $U \cap f^{-1}(R')$ consists of n distinct external rays, corresponding to the n distinct components of the set $\sqrt[n]{R} \subset \mathbb{C} \smallsetminus \mathbb{D}_r$. Each of these n

external rays R'_j will end at some solution z to the equation $f(z) = v$. But this equation has at least a double solution at the critical point c, so at least two of these external rays, say R'_1 and R'_2, will land at c. Evidently the union $R'_1 \cup R'_2 \subset \overline{U}$ will cut the plane into two connected open sets, which we will call V_0 and V_1.

Next note that each of the images $f(V_0)$ and $f(V_1)$ contains all points of the complex plane, except possibly for the points of R'. In fact each $f(V_k)$ is an open set. If $\hat{z} \in \mathbb{C}$ is a boundary point of $f(V_k)$, then we can choose a sequence of points $z_j \in V_k$ so that the images $f(z_j)$ converge to $\hat{z}$. The z_j must certainly be bounded, so we can choose a subsequence which converges to some point $z' \in \mathbb{C}$. Now $z' \notin V_k$ since $f(z') = \hat{z}$ is a boundary point and f is an open map, so it follows that $z' \in \partial V_k = R'_1 \cup R'_2$ and hence $\hat{z} \in R'$. Since $\mathbb{C} \smallsetminus R'$ is connected, this implies that

$$f(V_k) \supset \mathbb{C} \smallsetminus R' \supset K.$$

Now let $J_0 = J \cap V_0$ and $J_1 = J \cap V_1$. Then it follows that

$$f(J_0) = f(J_1) = J.$$

Note that J_0 and J_1 are disjoint compact sets with $J_0 \cup J_1 = J$. Similarly, we can split each J_k into two disjoint compact subsets $J_{k0} = J_k \cap f^{-1}(J_0)$ and $J_{k1} = J_k \cap f^{-1}(J_1)$, with $f(J_{k\ell}) = J_\ell$. Continuing inductively, we split J into 2^{p+1} disjoint compact sets

$$J_{k_0 \cdots k_p} = J_{k_0} \cap f^{-1}(J_{k_1}) \cap \cdots \cap f^{-p}(J_p),$$

with $f(J_{k_0 \cdots k_p}) = J_{k_1 \cdots k_p}$. Similarly, for any infinite sequence $k_0 \, k_1 \, k_2 \ldots$ of zeros and ones let $J_{k_0 k_1 k_2 \cdots}$ be the intersection of the nested sequence

$$J_{k_0} \supset J_{k_0 k_1} \supset J_{k_0 k_1 k_2} \supset \cdots.$$

Each such intersection is compact and non-vacuous. In this way, we obtain uncountably many disjoint non-vacuous subsets with union J. Every connected component of J must be contained in exactly one of these, so J has uncountably many components. The proof for the filled Julia set is completely analogous. □

The Green's function.

As in 9.2, the function $z \mapsto |\Phi(z)|$ extends continuously throughout the attracting basin $\mathbb{C} \smallsetminus K$, taking values $|\Phi(z)| > 1$. (This function is finite valued, since a polynomial has no poles in the finite plane.) In practice it is customary to work with the logarithm of $|\Phi|$.

9.6. Definition: By the *Green's function* or the *canonical potential function* associated with K we mean the function $G : \mathbb{C} \to [0, \infty)$ which is identically zero on K, and takes the values

$$G(z) \;=\; \log |\Phi(z)| \;=\; \lim_{k \to \infty} \frac{1}{n^k} \log |f^{\circ k}(z)| \;>\; 0$$

outside of K. It is not difficult to check that G is continuous everywhere and harmonic,

$$G_{xx} + G_{yy} \;=\; 0 \;,$$

outside of the Julia set. (Problem 9-b. Here the subscripts denote partial derivatives, where $z = x + iy$.) The curves $G = \text{constant} > 0$ in $\mathbb{C} \smallsetminus K$ are known as *equipotentials*. Note the equation

$$G(f(z)) \;=\; n\,G(z) \;,$$

which shows that f maps each equipotential to an equipotential.

Concluding problems.

Problem 9-a. Grand orbit closures. Let f be a rational function, and let $\mathcal{A}$ be the attracting basin of some superattracting fixed point. For any $z_0 \in \mathcal{A}$ show that the closure of the grand orbit of z_0 consists all points z with $|\phi(z)| = |\phi(z_0)|$, together with all iterated forward and backward images of such points, and (unless this set is a singleton) together with the Julia set of f. By way of contrast, in the case of a geometrically attracting basin, show that such a grand orbit closure consists of isolated points of $\mathcal{A}$, together with the grand orbit of the attracting point and together with the Julia set.

Problem 9-b. Harmonic functions. If U is a simply connected open set of complex numbers $z = x + iy$, show that a smooth function $G : U \to \mathbb{R}$ is harmonic, $G_{xx} + G_{yy} = 0$, if and only if there is a *conjugate* harmonic function $H : U \to \mathbb{R}$, uniquely defined up to an additive constant, satisfying

$$H_x \;=\; -G_y, \quad H_y \;=\; G_x \;,$$

so that $G + iH$ is holomorphic. For an arbitrary Riemann surface S, show that there is a corresponding concept of harmonic function $S \to \mathbb{R}$ which is independent of any choice of local uniformizing parameters. Show that a harmonic function cannot have any local maximum or minimum, unless it is constant. Show that any bounded harmonic function on the punctured disk $\mathbb{D} \smallsetminus \{0\}$ extends to a harmonic function on $\mathbb{D}$.

Now consider a polynomial f of degree $n \geq 2$. Show that the Green's function $G(z) = \log |\Phi(z)|$ is harmonic on $\mathbb{C} \smallsetminus K$, that it tends to zero as

z approaches K, and that it satisfies

$$G(z) = \log|z| + O(1) \qquad \text{as} \qquad |z| \to \infty.$$

(In other words, $G(z) - \log|z|$ is bounded for large $|z|$. A more precise estimate would be $G(z) = \log|z| + \log|a_n|/(n-1) + o(1)$ as $|z| \to \infty$, where a_n is the leading coefficient.) Show that the function G is uniquely characterized by these properties. Hence G is completely determined by the compact set $K = K(f)$, although our construction of G depends explicitly on the polynomial f.

Problem 9-c. Cellular sets and Riemann-Hurwitz. Here is another approach to Theorem 9.5. Again let f be a polynomial of degree $n \geq 2$. For each number $g > 0$ let V_g be the bounded open set consisting of all complex numbers z with $G(z) < g$. Using the maximum modulus principle, show that each connected component of V_g is simply connected. Hence the Euler characteristic $\chi(V_g)$ can be identified with the number of connected components of V_g. Show similarly that each component of V_g intersects the filled Julia set.

The Riemann-Hurwitz formula (7.2) applied to the map $f : V_g \to V_{ng}$ asserts that $n\chi(V_{ng}) - \chi(V_g)$ is equal to the number of critical points of f in V_g, counted with multiplicity. Since V_g is clearly connected for g sufficiently large, conclude that V_g is connected if and only if it contains all of the $n-1$ critical points of f.

A compact subset of Euclidean n-space is said to be *cellular*[*] if it is a nested intersection of closed topological n-cells, each containing the next in its interior. Show that the filled Julia set $K = \cap V_g$ is cellular (and hence connected) if and only if it contains all of the $n-1$ finite critical points of f. (In fact, if one of these critical points lies outside of K, and hence outside of some V_g, show that V_g and hence K is not connected.)

Problem 9-d. Quadratic polynomials. Now let $f(z) = z^2 + c$, and suppose that the critical orbit escapes to infinity. Let $V = V_{G(c)}$ be the open set consisting of all $z \in \mathbb{C}$ with $|\Phi(z)| < |\Phi(c)|$. Show that V is conformally isomorphic to $\mathbb{D}$, and that $f^{-1}(V)$ has two connected components. Conclude that $f^{-1}|_V$ has two holomorphic branches g_0 and g_1 mapping V into disjoint open subsets, each having compact closure in V. Show that each g_j strictly contracts the Poincaré metric of V. Proceeding as in Problem 4-e, show that J is a Cantor set, canonically

---

[*] Compare [Brown], where it is shown that a subset K of the sphere S^n is cellular if and only if its complement $S^n \setminus K$ is an open n-cell. This concept is of more interest in higher dimensions. In fact it is not hard to see that a compact subset of $\mathbb{C}$ is cellular if and only if it is connected with connected complement.

homeomorphic to the space of all infinite sequences $(j_0, j_1, j_2, \ldots)$ of zeros and ones.

§10. Parabolic Fixed Points: the Leau-Fatou Flower

Again we consider functions $f(z) = \lambda z + a_2 z^2 + a_3 z^3 + \cdots$ which are defined and holomorphic in some neighborhood of the origin, but in this section we suppose that the multiplier λ at the fixed point is a root of unity, $\lambda^q = 1$. Such a fixed point is said to be *parabolic*, provided that $f^{\circ q}$ is not the identity map. (Compare 4.4.) First consider the special case $\lambda = 1$. Then we can write our map as

$$f(z) = z(1 + a z^n + \text{(higher terms)}) = z + a z^{n+1} + \text{(higher terms)},$$
$$(10:1)$$

with $n \geq 1$ and $a \neq 0$. The integer $n + 1$ is called the *multiplicity* of the fixed point. We are concerned here with fixed points of multiplicity $n + 1 \geq 2$. (By definition, a fixed point has multiplicity equal to 1, or is *"simple"*, if and only if it has multiplier $\lambda \neq 1$, so that the graph of f intersects the diagonal transversally.)

By a "unit vector" at the origin we will mean simply a unit complex number $|v| = 1$, identified with the tangent vector to the smooth curve $t \mapsto t v$ at $t = 0$.

Definition. We will say that a unit vector v at the origin points in a *repelling direction* if $a v^n$ is real and positive, so that a vector from v to $v(1 + a v^n)$ points straight away from the origin. Similarly, it points in an *attracting direction* if $a v^n$ is real and negative.

Thus there are n equally spaced repelling directions at the origin, separated by n equally spaced attracting directions. Note that the repelling directions for f are just the attracting directions for the inverse map f^{-1}, which is also well defined and holomorphic in a neighborhood of the origin.

Here is a preliminary description of the local dynamics. Consider some orbit $z_0 \mapsto z_1 \mapsto \cdots$ for the map f of formula $(10:1)$. We will say that this orbit converges to zero *non-trivially* if $z_k \to 0$ as $k \to \infty$, but no z_k is actually equal to zero.

> **Lemma 10.1.** *If an orbit $f : z_0 \mapsto z_1 \mapsto \cdots$ converges to zero non-trivially, then the ratio $z_k/|z_k|$ must converge to some attracting unit vector as $k \to \infty$. Here any one of the n attracting unit vectors can occur. In fact for each $\epsilon > 0$ there is a $\delta > 0$ so that if the angle between $z_0/|z_0|$ and each of the n repelling vectors is greater than ϵ, and if $|z_0|$ is less than δ, then the orbit $z_0 \mapsto z_1 \mapsto \cdots$ must converge to zero non-trivially, and the associated ratios $z_k/|z_k|$ must converge to that attracting unit vector which is closest to $z_0/|z_0|$.*

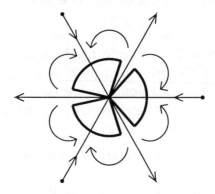

*Figure 14. Schematic picture of a parabolic point of multiplicity $n+$
$1 = 4$. (Here $a = -1$.) Each arrow indicates roughly how points
are moved by f. The three attracting directions are indicated by
arrows pointing towards the origin, and the three repelling directions
by arrows pointing away from the origin.*

For example, in Figure 14, if the three emphasized sectors have sufficiently small radius, then the Lemma says that any orbit $z_0 \mapsto z_1 \mapsto \cdots$
which starts out in one of these sectors will eventually converge to zero
within the sector, with $z_k/|z_k|$ converging to the associated attracting unit
vector.

Definition. If an orbit $z_0 \mapsto z_1 \mapsto \cdots$ under f converges to zero,
with $z_k/|z_k|$ tending to the attracting unit vector v, then we will say that
this orbit $\{z_k\}$ tends to zero *in the direction v*.

More generally, for any holomorphic map $f : S \to S$ on a Riemann
surface S and any fixed point $\hat{z}$ of multiplicity $n + 1 \geq 2$, it is easy
to generalize these constructions so that the following is true: *There are
exactly n distinct "attracting directions" in the tangent space to S at $\hat{z}$,
and any orbit which converges to $\hat{z}$ non-trivially must approach it in one
of these n attracting directions.*

10.2. Definition. Given such an attracting direction v_j in the tangent space of S at a multiple fixed point $\hat{z}$, the parabolic *basin of attraction*
$\mathcal{A}_j = \mathcal{A}(\hat{z}, v_j)$ is defined to be the set consisting of all $z_0 \in S$ for which the
orbit $z_0 \mapsto z_1 \mapsto \cdots$ converges to $\hat{z}$ in the direction v_j. Evidently these
basins $\mathcal{A}_1, \ldots, \mathcal{A}_n$ are disjoint fully invariant open sets, with the property
that an orbit $z_0 \mapsto z_1 \mapsto \cdots$ under f converges to $\hat{z}$ non-trivially if and
only if it belongs to one of the $\mathcal{A}_j$. The *immediate basin* $\mathcal{A}_j^0$ is defined to
be the unique connected component of $\mathcal{A}_j$ which maps into itself under f.

Figure 15. Julia set for $f(z) = z^5 + (.8 + .8i)z^4 + z$.   This map has a parabolic fixed point of rotation number zero and petal number three at $z = 0$ (and also an attracting fixed point at $z = -.8 - .8i$).   The immediate basins for the three attracting directions resemble balloons, pulled together at the parabolic point and separated by the three repelling directions.

Equivalently, we could define $\mathcal{A}_j$ to be that connected component of the Fatou set $S \smallsetminus J$ which contains z_k for large k , whenever $\{z_k\}$ converges to $\hat{z}$ in the direction v_j .

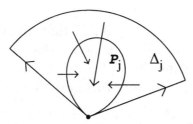

Figure 16. Sector Δ_j for the case $n = 3$, and an enclosed attracting petal $\mathcal{P}_j$.

Proof of 10.1. We will use the substitution $w = \phi(z) = c/z^n$, where $c = -1/(na)$. Suppose, to fix our ideas, that z ranges over the open sector Δ_j of radius ϵ and angle $2\pi/n$ consisting of all $re^{i\theta}v_j$ with $0 < r < \epsilon$ and $|\theta| < \pi/n$, where v_j is one of the n attracting directions. Then w will range correspondingly over the region $|w| > |c|/\epsilon^n$ in the slit plane $\mathbb{C} \smallsetminus (-\infty, 0]$. We will write

$$ f(z) = z(1 + az^n + o(z^n)) , $$

where the notation $o(z^n)$ stands for a remainder term, depending on z , which tends to zero faster than z^n so that $o(z^n)/z^n \to 0$ as $z \to 0$. The corresponding transformation in the w-plane is $w \mapsto F(w) = \phi \circ f \circ \phi^{-1}(w)$ where $\phi^{-1}(w) = \sqrt[n]{c/w}$, taking that branch of the n-th root with values in the sector Δ_j. Thus F is a well defined holomorphic map which is defined on the complement of a large disk in the slit plane, and takes values in $\mathbb{C}$. Note that

$$f \circ \phi^{-1}(w) = \sqrt[n]{c/w}\left(1 + a\frac{c}{w} + o\left(\frac{1}{w}\right)\right),$$

hence

$$F(w) = w\left(1 + a\frac{c}{w} + o\left(\frac{1}{w}\right)\right)^{-n} = w\left(1 + \frac{-nac}{w} + o\left(\frac{1}{w}\right)\right).$$

Since $nac = -1$, this can be written briefly as

$$F(w) = w + 1 + o(1),$$

where $o(1)$ stands for a remainder term which tends to zero as $|w| \to \infty$. In other words: For any $\eta > 0$ there exists an r_η so that

$$|F(w) - w - 1| < \eta \qquad \text{whenever} \qquad |w| > r_\eta. \qquad (10:2)$$

(We will later need the slightly more precise statement that

$$F(w) = w + 1 + O(1/\sqrt[n]{w}) \qquad \text{as} \qquad |w| \to \infty. \qquad (10:3)$$

That is, there exist constants r and C so that $|F(w) - w - 1| \leq C/\sqrt[n]{|w|}$ whenever $|w| \geq r$. This follows from a similar argument.)

In particular, it follows from $(10:2)$ that $\text{Re}(F(w)) > \text{Re}(w) + 1 - \eta$ when $|w| > r_\eta$. As an example, suppose that we choose $\eta = 1/2$, and choose w_0 in the half-plane $\text{Re}(w_0) > r_{1/2}$. Then clearly the orbit $w_0 \mapsto w_1 \mapsto \cdots$ under F will remain in this half-plane and satisfy

$$\text{Re}(w_k) > r_{1/2} + k/2, \qquad (10:4)$$

tending to infinity as $k \to \infty$. It then follows from $(10:2)$ that the difference $w_j - w_{j-1}$ tends to $+1$ as $j \to \infty$. Hence the ratio

$$\frac{w_k - w_0}{k} = \frac{1}{k}\sum_{1}^{k}(w_j - w_{j-1})$$

also tends to $+1$ as $k \to \infty$. Since w_0/k tends to 0, this proves that the sequence of complex numbers w_k is *asymptotic* to $+k$ as $k \to \infty$. *In particular, it follows that*

$$|w_k| \to \infty, \qquad \frac{w_k}{|w_k|} \to 1, \qquad \text{hence} \qquad z_k \to 0, \qquad \frac{z_k}{|z_k|} \to v_j,$$

as $k \to \infty$. (Just a little more work yields the more precise estimate $z_k \sim v_j \sqrt[n]{|c|/k}$ as $k \to \infty$.)

All of this goes through under the hypothesis that w_0 belongs to some half-plane $\mathrm{Re}(w_0) > r_\eta$, where $0 < \eta < 1$. However, we can weaken the hypothesis but obtain the same conclusion as follows. Draw two tangent half-lines from the circle of radius r_η centered at the origin, with slope $\pm\eta/\sqrt{1-\eta^2}$ as shown in Figure 17. Let U be the region to the right, bounded by these two half-lines and a circle arc. Then $|F(w) - w - 1| < \eta$ for all $w \in U$, and it follows by an easy geometric argument that F maps U into itself. Hence every orbit which starts in U eventually gets to the half-plane $\mathrm{Re}(w) > r_\eta$, and we see as above that $w_k \sim k$ as $k \to \infty$. Since η can be arbitrarily small, this proof takes care of all values of w_0 which are sufficiently far from the origin, as long as $w_0/|w_0|$ is bounded away from -1. That is, more precisely, every w_0 with $|w_0/|w_0| + 1| > \epsilon$, and with $|w_0|$ greater than some constant depending on ϵ must belong to the region $U = U_\eta$ for suitable choice of η. This completes the proof of 10.1. □

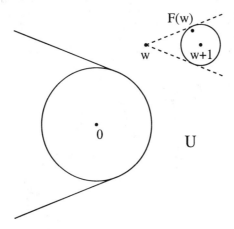

Figure 17. Diagram in the w-plane, showing circles of radius η about $w + 1$ and radius r_η about the origin, with lines of slope $\pm\eta/\sqrt{1-\eta^2}$.

More generally, if $\hat{z}$ is a periodic point of period k with multiplier $\lambda = e^{2\pi i p/q}$ for the map $f : S \to S$, then $\hat{z}$ is a fixed point of multiplier $+1$ for the iterate $f^{\circ kq}$. By definition, the parabolic basins for $f^{\circ kq}$ at $\hat{z}$ are also called parabolic basins for f.

10.3. Corollary. *For a holomorphic map $f : S \to S$, each parabolic basin $\mathcal{A}_j$ is contained in the Fatou set $S \smallsetminus J(f)$, but* each basin boundary $\partial \mathcal{A}_j$ is contained in the Julia set $J(f)$.

Proof. We already know by 4.4 that the fixed point $\hat{z}$ itself belongs to the Julia set. If an orbit $z_0 \mapsto z_1 \mapsto \cdots$ eventually lands at $\hat{z}$, or in other words converges "trivially" to $\hat{z}$, then it follows that z_0 belongs to the Julia set. Consider then a point $z_0 \in \partial \mathcal{A}_j$ whose orbit does not converge trivially to $\hat{z}$. Since z_0 is not in any of the attractive basins $\mathcal{A}_j$, the orbit $z_0 \mapsto z_1 \mapsto \cdots$ also cannot converge non-trivially to $\hat{z}$. Hence we can extract a subsequence $z_{k(i)}$ which is bounded away from $\hat{z}$. Since the sequence of iterates $f^{\circ k}$ converges to $\hat{z}$ throughout the open set $\mathcal{A}_j$, it follows that $\{f^{\circ k}\}$ can not be normal in any neighborhood of the boundary point z_0. The proof is now straightforward. $\square$

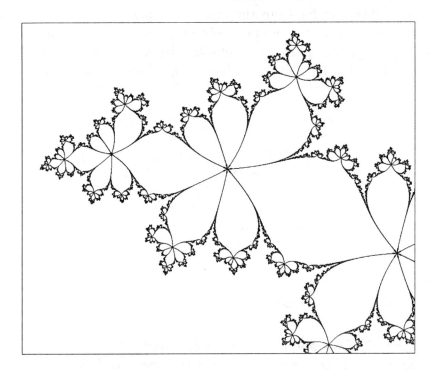

Figure 18. Julia set for $z \mapsto z^2 + e^{2\pi it}z$ with $t = 3/7$.

Now suppose that the multiplier λ at a fixed point is a q-th root of unity, say $\lambda = \exp(2\pi ip/q)$ where p/q is a fraction in lowest terms.

10.4. Lemma. *If the multiplier λ at a fixed point $f(\hat{z}) = \hat{z}$ is a primitive q-th root of unity, then the number n of attracting directions at $\hat{z}$ must be a multiple of q. In other words,* *the multiplicity* $n + 1$ *of* $\hat{z}$ *as a fixed point of* $f^{\circ q}$ *must be congruent to 1 modulo* q .

As an example, Figure 18 shows part of the Julia set for a quadratic map f having a fixed point of multiplier $\lambda = e^{2\pi i(3/7)}$ at the origin, near the center of the picture. In this case, the 7-fold iterate $f^{\circ 7}$ is a map of degree 128 with a fixed point of multiplicity $7 + 1 = 8$ at the origin. The seven immediate attracting basins are clearly visible in the Figure.

Proof of 10.4. If v is any attracting direction for $f^{\circ q}$ at $\hat{z}$, then we can choose an orbit $z_0 \mapsto z_q \mapsto z_{2q} \cdots$ under $f^{\circ q}$ which converges to $\hat{z}$ in the direction v . Evidently the image $z_1 \mapsto z_{q+1} \mapsto z_{2q+1} \mapsto \cdots$ under f will be an orbit which converges to $\hat{z}$ in the direction λv . Thus multiplication by $\lambda = e^{2\pi i p/q}$ permutes the n attracting directions, and the conclusion follows easily. □

Remark. If we replace $f = f_0$ by a nearby map f_t , so as to change λ slightly, then the $(n + 1)$-fold fixed point $\hat{z}$ of $f^{\circ q}$ will split up into $n + 1$ simple fixed points of $f_t^{\circ q}$. Since $\hat{z}$ is a simple fixed point of $f^{\circ k}$ for $k < q$, it follows that only one of these $n + 1$ points will be fixed by f_t or by $f_t^{\circ k}$. The remaining n will partition into n/q orbits, each of period exactly q .

We return to the case $\lambda = 1$. It is often convenient to have a purely local analogue for the global concept of "basin of attraction". Let $\hat{z}$ be a fixed point of multiplicity $n + 1 \geq 2$. Choose a neighborhood N of $\hat{z}$ which is small enough so that f maps N diffeomorphically onto some neighborhood N' of $\hat{z}$. Thus the inverse function f^{-1} is uniquely defined and holomorphic as a map from N' onto N . Let v be an attracting direction at $\hat{z}$.

Definition. A simply connected open set $\mathcal{P} \subset N \cap N'$, with $f(\mathcal{P}) \subset \mathcal{P}$, will be called an *attracting petal* for f in the direction v at $\hat{z}$ if:

(1) the sequence of iterates $f^{\circ k}$ restricted to $\mathcal{P}$ converges uniformly to the constant function $z \mapsto \hat{z}$, and

(2) an orbit $z_0 \mapsto z_1 \mapsto \cdots$ under f is eventually absorbed by this set $\mathcal{P}$ if and only if it converges to $\hat{z}$ in the direction v .

Similarly, a simply connected open set $\mathcal{P}' \subset N \cap N'$ is a *repelling petal* for the repelling direction v' if $\mathcal{P}'$ is an attracting petal for f^{-1} in this direction.

10.5. Leau-Fatou Flower Theorem. *If* $\hat{z}$ *is a fixed point of multiplicity* $n + 1 \geq 2$, *then there exist attracting petals* $\mathcal{P}_1 \ldots, \mathcal{P}_n$ *for the* n *attracting directions at* $\hat{z}$, *and repelling*

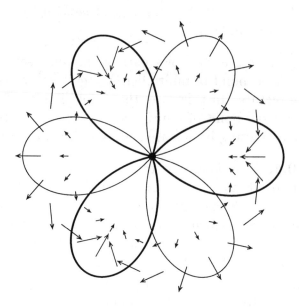

Figure 19. Leau-Fatou Flower with three attracting petals and three repelling petals. Each arrow points from z to $f(z)$.

petals $\mathcal{P}'_1, \ldots, \mathcal{P}'_n$ for the n repelling directions, so that the union of these $2n$ petals, together with $\hat{z}$ itself, forms a neighborhood N_0 of $\hat{z}$. Furthermore, these $2n$ petals are arranged cyclically, as illustrated in Figure 19, so that two petals intersect if and only if the angle between their central directions is π/n.

Thus each attracting petal intersects exactly two repelling petals, if $n > 1$. (In the special case $n = 1$, there is just one attracting petal and one repelling petal, and their intersection $\mathcal{P} \cap \mathcal{P}'$ has two distinct connected components.)

The proof of 10.5 follows easily from the discussion above. The open set U in the w-plane corresponds under the change of coordinates $w = c/z^n$ to any one of the required attracting petals. Details will be left to the reader. □

If $f : S \to S$ is a globally defined holomorphic function, and $\hat{z}$ is a fixed point of multiplicity $n+1 \geq 2$, then each attracting petal $\mathcal{P}_j$ about $\hat{z}$ determines a corresponding parabolic *basin of attraction* $\mathcal{A}_j$, consists of all z_0 for which the orbit $z_0 \mapsto z_1 \mapsto \cdots$ eventually lands in $\mathcal{P}_j$, and hence converges to the fixed point in the associated direction v_j. (Compare 10.2.)

We can further describe the geometry around a parabolic fixed point as follows. As in (10 : 1), we consider a local analytic map with a fixed point of multiplier $\lambda = 1$. Let $\mathcal{P}$ be either one of the n attracting petals or one of the n repelling petals, as described in the Flower Theorem 10.5. Form an identification space $\mathcal{P}/f$ from $\mathcal{P}$ by identifying z with $f(z)$ whenever both z and $f(z)$ belong to $\mathcal{P}$. (This means that z is identified with $f(z)$ for every $z \in \mathcal{P}$ in the case of an attracting petal, and for every $z \in \mathcal{P} \cap f^{-1}(\mathcal{P})$ in the case of a repelling petal.)

10.6. Cylinder Theorem. *For each attracting or repelling petal $\mathcal{P}$, the quotient manifold $\mathcal{P}/f$ is conformally isomorphic to the infinite cylinder $\mathbb{C}/\mathbb{Z}$.*

By definition, the quotient $\mathcal{P}/f$ is called an *Écalle cylinder* for $\mathcal{P}$. This term is due to Douady, suggested by the work of Écalle on holomorphic maps tangent to the identity. The behavior of Écalle cylinders under perturbation of the mapping f is a very important topic in holomorphic dynamics. (Compare Lavaurs, Shishikura.)

We will derive 10.6 as an immediate consequence of the following basic result, which was proved by Leau and Fatou. However, it would be equally possible to first prove 10.6 and then derive 10.7. (Compare Problem 10-c.)

10.7. Parabolic Linearization Theorem. *There is one, and up to composition with a translation only one, conformal embedding α from $\mathcal{P}$ into the universal covering space $\mathbb{C}$ of the cylinder which satisfies the Abel functional equation*

$$\alpha(f(z)) \;=\; 1 + \alpha(z)$$

for all $z \in \mathcal{P} \cap f^{-1}(\mathcal{P})$. With suitable choice of $\mathcal{P}$, the image $\alpha(\mathcal{P}) \subset \mathbb{C}$ will contain some right half-plane $\{w; \mathrm{Re}(w) > c\}$ in the case of an attracting petal, or some left half-plane in the case of a repelling petal.

The linearizing coordinate $\alpha(z)$ is often referred to as a *Fatou coordinate* in $\mathcal{P}$.

Proof of 10.7 and 10.6. The following argument, loosely following Steinmetz, is completely constructive, and can be used for actual computation of Fatou coordinates, although convergence is rather slow. It suffices to consider the case of an attracting petal. As in the proof of 10.1, we will work with the coordinate $w = c/z^n$ and the function $F : U \to U$ which is holomorphically conjugate to $f : \mathcal{P} \to \mathcal{P}$. According to the estimate (10 : 3), this satisfies

$$F(w) \;=\; w + 1 + O(1/|w|^{\epsilon}) \quad \text{as} \quad |w| \to \infty, \tag{10 : 3$'$}$$

where $\epsilon = 1/n > 0$. Using this, we will first prove the following estimate for the first derivative of F,

$$F'(w) = 1 + O\left(\frac{1}{|w|^{1+\epsilon}}\right) \qquad \text{as} \qquad |w| \to \infty.$$

More explicitly, choose constants R and C so that $|F(w) - w - 1| < C/|w|^\epsilon$ for $|w| > R/2$. Then

$$|F'(w) - 1| < C\,|2/w|^{1+\epsilon} \qquad \text{for} \qquad |w| > R.$$

This follows since the map $w \mapsto F(w) - w - 1$, restricted to a disk of radius $|w_0/2|$ centered at w_0, takes values in a disk of radius $C/|w_0/2|^\epsilon$ whenever $|w_0| > R$. Hence its derivative at w_0 is bounded by the ratio of these two numbers by Lemma 1.2'.

Now suppose that w and $\hat{w}$ are any two points in the right half-plane $\mathrm{Re}(w) \geq R$. Averaging the expression $F'(w) - 1$ over the line segment from w to $\hat{w}$, we see that

$$\left|\frac{F(w) - F(\hat{w})}{w - \hat{w}} - 1\right| \leq C\,(2/\xi)^{1+\epsilon},$$

where $\xi \geq R$ is the smaller of the two numbers $\mathrm{Re}(w)$ and $\mathrm{Re}(\hat{w})$. In particular, given any two orbits $w_0 \mapsto w_1 \mapsto \cdots$ and $\hat{w}_0 \mapsto \hat{w}_1 \mapsto \cdots$ in this half-plane, since we know by $(10:4)$ that $\mathrm{Re}(w_k)$ and $\mathrm{Re}(\hat{w}_k)$ increase at least linearly with k, we have

$$\left|\frac{w_k - \hat{w}_k}{w_{k-1} - \hat{w}_{k-1}} - 1\right| \leq \frac{C'}{k^{1+\epsilon}} \qquad \text{hence} \qquad \left|\frac{w_k - \hat{w}_k}{w_{k-1} - \hat{w}_{k-1}}\right| \leq 1 + \frac{C'}{k^{1+\epsilon}},$$

$$(10:5)$$

for suitable choice of C'. Note that the infinite product

$$P = \prod_{j \geq 1} \left(1 + C'/j^{1+\epsilon}\right)$$

is finite. Using the right side of $(10:5)$ and taking a suitable finite product, it follows that

$$|w_{k-1} - \hat{w}_{k-1}| \leq |w_0 - \hat{w}_0|\,P$$

for every k. Multiplying this inequality by the left side of $(10:5)$, it follows that

$$|(w_k - \hat{w}_k) - (w_{k-1} - \hat{w}_{k-1})| \leq PC'\,|w_0 - \hat{w}_0|/k^{1+\epsilon}.$$

Summing over k, since $\sum 1/k^{1+\epsilon} < \infty$, we see that the limit

$$\lim_{k \to \infty} (w_k - \hat{w}_k) = (w_0 - \hat{w}_0) + \sum_{k \geq 1} ((w_k - \hat{w}_k) - (w_{k-1} - \hat{w}_{k-1})) \quad (10:6)$$

actually exists.

Now set $\beta(w_0)$ equal to this limit $(10:6)$, or in other words set

$$\beta(w) \;=\; \lim_{k\to\infty}\left(F^{\circ k}(w) - F^{\circ k}(\hat{w}_0)\right),$$

where $\hat{w}_0$ is some fixed base point in the half-plane. The proof shows that this convergence is uniform on compact sets, hence the limit function $w \mapsto \beta(w)$ is holomorphic. Next note that

$$\beta(F(w)) \;=\; \lim\left(w_{k+1} - \hat{w}_k\right) \;=\; \lim\left(w_{k+1} - \hat{w}_{k+1}\right) + \lim\left(\hat{w}_{k+1} - \hat{w}_k\right),$$

where the two expressions on the right converge to $\beta(w)$ and to $+1$ respectively. This shows that β satisfies the Abel equation $\beta(F(w)) = \beta(w) + 1$ for all w in the half-plane $\mathrm{Re}(w) > R$. It follows that the composition $\alpha(z) = \beta(c/z^n)$ is defined and satisfies the required Abel equation

$$\alpha(f(z)) \;=\; \alpha(z) + 1$$

for all z in a corresponding attracting petal $\mathcal{P}_R$. We can extend to a completely arbitrary attracting petal $\mathcal{P}$, simply by setting $\alpha(z)$ equal to $\alpha(f^{\circ k}(z)) - k$, choosing k is large enough so that $f^{\circ k}(z) \in \mathcal{P}_R$.

If we divide by $w_0 - \hat{w}_0$, then it follows easily from the argument above that the convergence

$$\frac{w_k - \hat{w}_k}{w_0 - \hat{w}_0} \;\to\; \frac{\beta(w_0)}{w_0 - \hat{w}_0} \qquad \text{as} \qquad k \to \infty$$

is uniform, as w_0 varies throughout the entire right half-plane. Suppose that we keep k and $\hat{w}_0$ fixed, but let $|w_0|$ tend to infinity. Then the difference $w_k - w_0$ converges to k and hence remains bounded. Therefore the ratio on the left tends to $+1$. Thus we obtain the asymptotic equality $\beta(w) \sim w - \hat{w}_0$ hence

$$\beta(w) \;\sim\; w, \qquad\qquad\qquad (10:7)$$

as $|w|$ tends to infinity with $\mathrm{Re}(w) \geq R$. Making use of Rouché's theorem, it follows that the image of the right half-plane under β (and therefore the image of the corresponding petal $\mathcal{P}_R$ under α), contains some entire right half-plane.

In order to prove that the function β is injective, we will use the inequality

$$\left|\frac{w_k - \hat{w}_k}{w_{k-1} - \hat{w}_{k-1}}\right| \;\geq\; 1 - \frac{C'}{k^{1+\epsilon}},$$

which follows from the left half of $(10:5)$. Since $\prod_N^\infty(1 - C'/k^{1+\epsilon}) > 0$, for N sufficiently large, it follows that $\beta(w_0) \neq \beta(\hat{w}_0)$ whenever $w_0 \neq \hat{w}_0$. (Here we are making use of the fact that the map F itself is known to be univalent, so that $w_N \neq \hat{w}_N$.) Since the choice of basepoint is arbitrary, this proves that β is injective, and therefore that the derivative $\beta'(w)$ is never zero. Thus β is univalent, and it follows similarly that α is univalent.

The conformal isomorphism $\mathcal{P}/f \overset{\cong}{\longrightarrow} \mathbb{C}/\mathbb{Z}$ of 10.6 can now be obtained by mapping the equivalence class $\{z_0, z_1, z_2, \ldots\}$ to the residue class of $\alpha(z_0)$ modulo $\mathbb{Z}$. To see that α is unique up to translation, it is convenient to use the conformal equivalence $\mathbb{C}/\mathbb{Z} \cong \mathbb{C} \smallsetminus \{0\}$. But any conformal equivalence from $\mathbb{C} \smallsetminus \{0\}$ to itself extends uniquely to a conformal equivalence of the Riemann sphere, which must have the form $z \mapsto cz$ or $z \mapsto c/z$ for some constant $c \neq 0$. The first case corresponds to a translation of $\mathbb{C}/\mathbb{Z}$, while it is easy to see that the second case cannot occur. This completes the proof of 10.7 and 10.6. $\quad\square$

Remark 10.8. Note that this preferred Fatou coordinate system is defined only within one of the $2n$ attracting or repelling petals. In order to describe a full neighborhood of the parabolic fixed point, we would have to describe how these $2n$ Fatou coordinate systems are to be pasted together in pairs, by means of univalent mappings which satisfy the functional equation $\psi(\alpha + 1) = \psi(\alpha) + 1$ for α in the overlap region. In fact each of the $2n$ required pasting maps can be extended to a map which is defined either in some upper half-plane $\operatorname{Im}(\alpha) > \text{constant}$ or alternately in some lower half-plane $\operatorname{Im}(\alpha) < \text{constant}$. Such a map necessarily has the form $\psi(\alpha) = \alpha + \Psi(e^{\pm 2\pi i \alpha})$, where Ψ can be extended to a map which is defined and holomorphic in some neighborhood of the origin, and where $\pm$ is the sign of $\operatorname{Im}(\alpha)$, so that $|e^{\pm 2\pi i \alpha}| < 1$. These maps are not quite uniquely defined, since the Fatou coordinates are only defined up to an additive constant. However, each of the power series Ψ depends on infinitely many parameters. It this way, one sees that *there can be no normal form depending on only finitely many parameters for a general holomorphic map f in the neighborhood of a parabolic fixed point.* (Compare Malgrange, Voronin.) On the other hand, if we allow a change of coordinate given by a formal power series, then there is a normal form $z \mapsto z + z^{n+1} + \beta \, z^{2n+1}$ depending on just one complex parameter (Problem 10-d), while if we allow a topological change of coordinate then Camacho showed that the normal form $z \mapsto z + z^{n+1}$ will suffice.

Although the coordinate α is well defined only up to an additive constant, its differential $d\alpha$ is uniquely defined. Thus another way of describing 10.7 is to say that within each attracting or repelling petal there is a unique holomorphic 1-form $d\alpha = (d\alpha/dz)\, dz$ which is f-invariant, and satisfies $\int_z^{f(z)} d\alpha = +1$. Equivalently, within each petal there is a uniquely defined holomorphic vector field

$$\frac{dz}{d\alpha}\frac{\partial}{\partial z}$$

with the following property: *The time one map for the solutions of the associated differential equation*

$$\frac{dz(t)}{dt} = \frac{1}{d\alpha(z)/dz}$$

is precisely the given map f . In particular, we can write f as an iterate $f = g \circ g$ within each petal, where g is the time $1/2$ map for this differential equation. However, as in the paragraph above, there is no reason to expect these differential forms or vector fields or functions $g = f^{\circ 1/2}$ to match properly on the overlap between two adjacent petals. If we consider f only as a formal power series $z + az^{n+1} + \cdots$, then the equation $f = g \circ g$ has a unique formal power series solution $g(z) = z + (a/2)z^{n+1} + \cdots$. However, there is no reason to expect this new formal power series to converge.

Now suppose that $f : S \to S$ is a globally defined holomorphic map. Although attracting petals behave much like repelling petals in the local theory, they behave quite differently in the large.

10.9. Corollary. *If* $\mathcal{P} \subset S$ *is an attracting petal for* f *, then the Fatou map*

$$\alpha : \mathcal{P} \to \mathbb{C}$$

extends uniquely to a map $\mathcal{A} \to \mathbb{C}$ *which is defined and holomorphic throughout the attractive basin of* $\mathcal{P}$ *, still satisfying the Abel equation* $\alpha(f(z)) = 1 + \alpha(z)$.

In the case of a repelling petal, the analogous statement is the following.

10.10. Corollary. *If* $\mathcal{P}'$ *is a repelling petal for* $f : S \to S$ *, then the inverse map*

$$\alpha^{-1} : \alpha(\mathcal{P}') \to \mathcal{P}'$$

extends uniquely to a globally defined holomorphic map $\gamma : \mathbb{C} \to S$ *which satisfies the corresponding equation*

$$f(\gamma(w)) = \gamma(1 + w) .$$

The proofs are completely analogous to the corresponding proofs in 8.4 and 8.10. □

In the case of a non-linear rational map $f : \widehat{\mathbb{C}} \to \widehat{\mathbb{C}}$, the extended map of 10.9 is surjective. However, it is not univalent, but rather has critical points whenever some iterate $f \circ \cdots \circ f$ has a critical point. Note the following basic result.

>   **10.11. Corollary.** *If $\hat{z}$ is a parabolic fixed point with multiplier $\lambda = 1$ for a rational map, then each immediate basin for $\hat{z}$ contains at least one critical point of f. Furthermore, each basin contains one and only one petal $\mathcal{P}^*$ which maps univalently onto some right half-plane under α and which is maximal with respect to this property. This preferred petal $\mathcal{P}^*$ always has one or more critical points on its boundary.*

The proof is completely analogous to the corresponding proof in 8.5. It is not difficult to show that α^{-1} can be defined throughout *some* right half-plane. If we try to extend leftwards by analytic continuation then we must run into an obstruction, which can only be a critical point of f. (For an alternative proof that every parabolic basin contains a critical point, see Milnor & Thurston, pp. 512-515.) $\square$

As an example, Figure 20 illustrates the map $f(z) = z^2 + z$, with a parabolic fixed point of multiplier $\lambda = 1$ at $z = 0$, which is the cusp point at the right center of the picture. Here the Julia set J is the outer Jordan curve (the "cauliflower") bounding the basin of attraction $\mathcal{A}$. The critical point $w = -1/2$ lies exactly at the center of symmetry. All orbits in this basin $\mathcal{A}$ converge towards $z = 0$ to the right. The curves $\mathrm{Re}(\alpha(z)) = $ constant $\in \mathbb{Z}$ have been drawn in, using the normalization $\alpha(w) = 0$. Thus the preferred petal $\mathcal{P}^*$, with the critical point on $\partial \mathcal{P}^*$, is bounded by the right half of the central figure-∞ shaped curve. Note that the function $z \mapsto \mathrm{Re}(\alpha(z))$ has a saddle critical point at each iterated preimage of w. This function $\mathrm{Re}(\alpha(z))$ oscillates wildly as z tends to $J = \partial \mathcal{A}$.

As an immediate consequence of 10.11 we have the following.

>   **10.12. Corollary.** *A rational map can have at most finitely many parabolic periodic points. In fact for a map of degree $d \geq 2$ the number of parabolic cycles plus the number of attracting cycles is at most $2d - 2$.*

More precisely, the number of cycles of Fatou components which are either immediate parabolic basins or immediate attracting basins is at most equal to the number of distinct critical points. The proof is essentially the same as that given in 8.6. These attracting and parabolic basins must be disjoint, and each cycle of basins must contain at least one critical point. $\square$

(For sharper results, see Shishikura (1987), as well as A. Epstein.)

Figure 20. Julia set for $z \mapsto z^2 + z$, with the curves $\mathrm{Re}(\alpha(z)) \in \mathbb{Z}$ drawn in.

Problem 10-a. Repelling petals and the Julia set. If f is a non-linear rational function, show that every repelling petal must intersect the Julia set of f .

Problem 10-b. No small cycles. With notation as in 10.5, show that no orbit $z_0 \mapsto z_1 \mapsto \cdots$ under f can be contained in the union $\mathcal{P}'_1 \cup \cdots \cup \mathcal{P}'_n$ of repelling petals. Conclude that the only periodic orbit which is completely contained in the neighborhood

$$N_0 = \{\hat{z}\} \cup \mathcal{P}_1 \cup \cdots \cup \mathcal{P}_n \cup \mathcal{P}'_1 \cup \cdots \cup \mathcal{P}'_n$$

is the fixed point $\hat{z}$ itself. On the other hand, show that any non-linear rational function has orbits which return to every repelling petal infinitely often. (Compare 4.13. Using §14, one can also show that there are periodic points arbitrarily close to $\hat{z}$ in each $\mathcal{P}'_j$.)

Problem 10-c. Alternate Proof of 10.6. Suppose that g maps some right half-plane into itself with $|g(w) - w - 1| < 1/2$. After restricting to a smaller right half-plane $\bar{H}_c = \{w : \mathrm{Re}(w) \geq c\}$, show using 1.3 that $|g'(w) - 1| < 1/3$ say, and that g is univalent. Since the quotient surface $S = \bar{H}_c/g$ has free cyclic fundamental group, it must be conformally isomorphic either to an annulus, punctured disk, or cylinder. (Problem 2-g.) If S were isomorphic to the annulus $A_r = \{z : 1 < |z| < r\}$, then it would have a flat metric $|dz/z|$ of finite area, such that each loop $|z| = \mathrm{constant}$ has fixed length 2π. Correspondingly, the half-plane $\bar{H}_c$ would have a conformal metric $\gamma(w)|dw|$ of finite area such that the geodesic distance from w to $g(w)$ is exactly 2π. Hence the integral $\int \gamma |dz|$ along a straight line from w to $g(w)$ would be at least 2π. Using the Schwarz inequality $(\int 1 \cdot \gamma dt)^2 \leq (\int 1\, dt)(\int \gamma^2 dt)$, and using the non-holomorphic coordinate system

$$(t,\, \eta) \;\mapsto\; w \;=\; (1-t)(c+i\eta) + tg(c+i\eta)\,,$$

show that the area

$$\int\!\!\int \gamma^2 du\, dv \;=\; \int_{-\infty}^{\infty}\!\!\int_0^1 \gamma^2 \frac{\partial(u,v)}{\partial(t,\eta)} dt\, d\eta$$

of a fundamental domain $0 \leq t \leq 1$ for g on $\bar{H}_c$ must be infinite, thus yielding a contradiction. Similarly, since each of the half fundamental domains $\eta \geq 0$ and $\eta \leq 0$ has infinite area, show that S cannot be a punctured disk.

Problem 10-d. A formal normal form. Suppose that f is given by a power series of the form

$$f(z) \;=\; z + z^m + \text{(higher terms)} \qquad\qquad (10:8)$$

with $m \geq 2$, and let g be a local diffeomorphism of the form $g(z) = z + cz^k$ with $k \geq 2$. Show that

$$f(g(z)) - g(f(z)) \;=\; (m-k)\, c\, z^{m+k-1} + \text{(higher terms)}\,,$$

or equivalently that

$$g^{-1} \circ f \circ g(z) \;=\; f(z) + (m-k)\, c\, z^{m+k-1} + \text{(higher terms)}\,.$$

Conclude inductively that by such conjugations we can eliminate terms of any degree other than 1, m and $2m-1$ from the power series for f. Thus f is locally holomorphically conjugate to a map of the form

$$g(z) \;=\; z + a\, z^m + b\, z^{2m-1} + \text{(terms of degree} > N)\,,$$

where N can be arbitrarily large. Conclude that conjugation by a possibly non-convergent formal power series $\psi(z) = z + c_2\, z^2 + c_3\, z^3 + \cdots$ can transform any f of the form $(10:8)$ into the normal form

$$z \;\mapsto\; z + z^m + b\, z^{2m-1}\,.$$

Remarks. This coefficient b is a conjugacy class invariant, so that no further simplification is possible. (Compare Problem 12-b.) The power series ψ is definitely not convergent in general. In fact, as noted in 10.8, it would take infinitely many complex parameters to specify the map f up to local holomorphic conjugacy. (Compare [Malgrange], [Voronin].)

Problem 10-e. Immediate parabolic basins. By an argument similar to that of 8.7, show that the complement of an immediate parabolic basin is either connected or else has uncountably many connected components. (However, compare Problem 10-f(4).)

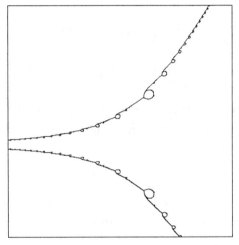

Figure 21. Julia set for $z \mapsto z + 1/(1 + z^2)$. (Problem 10-f(4).)

Problem 10-f. Examples with a parabolic point at infinity. (Compare Milnor 1993, §8.)

(1) For $f(z) = z - 1/z$ show that there is a parabolic point at infinity with two attracting directions. Since the upper and lower half-planes map into themselves, conclude that $J = \mathbb{R} \cup \{\infty\}$. (For a similar non-parabolic example, see Problem 7-a.)

(2) For $f(z) = z - 1/z + 1$ show that J is a Cantor set contained in $\mathbb{R} \cup \{\infty\}$.

(3) For $f(z) = z + 1/z - 2$ show that J is the interval $[0, +\infty]$.

(4) For $f(z) = z + 1/(1 + z^2)$ show that there are three attracting directions at infinity. Show that one of the three immediate parabolic basins contains all of $\mathbb{R}$, and hence *nearly* disconnects the Riemann sphere.

§11. Cremer Points and Siegel Disks

Once more we consider maps of the form

$$f(z) = \lambda z + a_2 z^2 + a_3 z^3 + \cdots ,$$

which are defined and holomorphic throughout some neighborhood of the origin, with a fixed point of multiplier λ at the origin. In §8 and §9 we supposed that $|\lambda| \neq 1$, while in §10 we took λ to be a root of unity. This section considers the remaining cases where $|\lambda| = 1$ but λ is not a root of unity. Thus we assume that the multiplier λ can be written as

$$\lambda = e^{2\pi i \xi} \qquad \text{with } \xi \text{ real and irrational} .$$

Briefly, we will say that the origin is an *irrationally indifferent* fixed point. The number $\xi \in \mathbb{R}/\mathbb{Z}$ is called the *rotation number* for the tangent space at the fixed point. (Remark: A theorem of [Naĭshul'] asserts that this rotation number is a local topological invariant.)

The fundamental question here is whether or not there exists a local change of coordinate $z = h(w)$ which conjugates f to the irrational rotation $w \mapsto \lambda w$, so that

$$f(h(w)) = h(\lambda w)$$

near the origin. (Compare 8.2.) In the special case of a globally defined rational function, the following is an immediate consequence of Theorem 5.2.

11.1. Lemma. *Let f be a rational function of degree two or more with a fixed point z_0 which is indifferent, $|f'(z_0)| = 1$. Then the following three conditions are equivalent to each other:*

- *f is locally linearizable around z_0*
- *z_0 belongs to the Fatou set $\widehat{\mathbb{C}} \smallsetminus J(f)$*
- *the connected component U of the Fatou set containing z_0 is conformally isomorphic to the unit disk under an isomorphism which conjugates f on U to multiplication by λ on the disk.*

Proof. If f is locally linearizable around z_0, then the iterates of f in a suitable neighborhood of z_0 correspond to iterated rotations of a small disk, and hence form a normal family. Thus z_0 belongs to the Fatou set. Conversely, whenever z_0 belongs to the Fatou set, we see from 5.2 that the entire Fatou component U of z_0 must be conformally isomorphic to the unit disk, with $f|_U$ conjugate to multiplication by λ on $\mathbb{D}$. $\square$

Definition. We will say that an irrationally indifferent fixed point is either a *Siegel point* or a *Cremer point* according as a local linearization is possible or not. A Fatou component on which f is conformally conjugate to a rotation of the unit disk is called a *Siegel disk* or a *rotation disk*, with *center* z_0. (In the classical literature, Siegel points were called "centers", and the question as to their existence was called the "center problem".)

This section will first survey what is known about the local linearization problem, and then prove some of the easier results. Finally, it will describe the relation between Cremer points or Siegel disks and the critical points of a rational map.

At the International Congress in 1912, E. Kasner conjectured that such a linearization is *always* possible. Five years later, G. A. Pfeiffer disproved this conjecture by giving a rather complicated description of certain holomorphic functions for which no local linearization is possible. In 1919 Julia claimed to settle the question completely for rational functions of degree two or more by showing that such a linearization is *never* possible; however, his proof was wrong. H. Cremer put the situation in much clearer perspective in 1927 with a result which we can state as follows.

11.2. Cremer Non-Linearization Theorem. *Given λ on the unit circle and given $d \geq 2$, if the d^q-th root of $1/|\lambda^q - 1|$ is unbounded as $q \to \infty$, then no fixed point of multiplier λ for a rational function of degree d can be locally linearizable.*

This will be proved below. It is convenient to say that a property of an angle $\xi \in \mathbb{R}/\mathbb{Z}$ is true for *generic* ξ if the set of ξ for which it is true contains a countable intersection of dense open subsets of $\mathbb{R}/\mathbb{Z}$. According to Baire, such a countable intersection of dense open sets is necessarily dense and uncountably infinite. (See Problem 4-j.)

11.3. Corollary. *For a generic choice of rotation number $\xi \in \mathbb{R}/\mathbb{Z}$, if z_0 is a fixed point of multiplier $e^{2\pi i \xi}$ for a completely arbitrary rational function f of degree two or more, then there is no local linearizing coordinate about z_0.*

(Compare Problem 11-b.) The question as to whether this statement is actually true for *all* ξ remained open until 1942, when Siegel proved the following. Again let $\lambda = e^{2\pi i \xi}$ with $\xi \in \mathbb{R} \setminus \mathbb{Q}$.

11.4. Siegel Linearization Theorem. *If $1/|\lambda^q - 1|$ is less than some polynomial function of q, then every germ of a holomorphic map with fixed point of multiplier λ is locally linearizable.*

11.5. Corollary. *For every ξ outside of a set of Lebesgue*

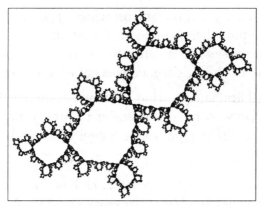

Figure 22a. Julia set for $z^2 + e^{2\pi i\xi}z$ with $\xi = \sqrt[3]{1/4}$
$= .62996\cdots$. The large region to the lower left is a Siegel disk.

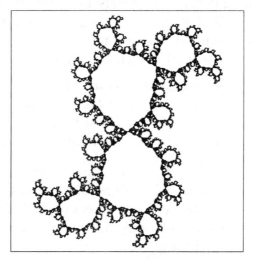

Figure 22b. Corresponding Julia set with a
randomly chosen angle $\xi = .7870595\cdots$

measure zero, we can conclude that every holomorphic germ with
a fixed point of multiplier $e^{2\pi i\xi}$ is locally linearizable.

In other words, if the angle $\xi \in \mathbb{R}/\mathbb{Z}$ is "randomly chosen" with respect to Lebesgue measure, then with probability one every rational function with a fixed point of multiplier $e^{2\pi i\xi}$ will have a corresponding Siegel disk. See Figure 22b for an example. These two results will not be proved in these notes. (However, one special case of 11.5 will be proved below, in 11.14.) For the proof of 11.4, see [Siegel] or [Siegel-Moser] or [Carleson-Gamelin] or [Zehnder], and for the implication 11.4 $\Rightarrow$ 11.5 see 11.7 below.

Remark. Comparing 11.3 and 11.5, we see that there is a total contrast between behavior for *generic* ξ and behavior for *almost every* ξ. This contrast is quite startling, but is not uncommon in dynamics. (Compare the discussion of the iterated exponential map in §6.) In applied dynamics, it is usually understood that behavior which occurs for a set of parameter values of measure zero has no importance, and can be ignored. However, even in applied dynamics the study of generic behavior remains an extremely valuable tool.

In order to understand these statements, as well as sharper results which have been obtained more recently, it is convenient to introduce a number of different classes of irrational numbers, which are related to each other as indicated schematically in Figure 23.

Let κ be a positive real number. By definition, an irrational number ξ is said to be *Diophantine of order* κ if there exists $\epsilon > 0$ so that

$$\left| \xi - \frac{p}{q} \right| > \frac{\epsilon}{q^\kappa} \qquad \text{for every rational number } p/q . \qquad (11:1)$$

The class of all such numbers will be denoted by $\mathcal{D}(\kappa)$. Setting $\lambda = e^{2\pi i \xi}$ as above, if p is the closest integer to $q\xi$ so that $|q\xi - p| \le 1/2$, note the order of magnitude estimate

$$|\lambda^q - 1| = |2 \sin (\pi(q\xi - p))| \asymp 2\pi |q\xi - p| .$$

(More precisely, it is not hard to see that $4|q\xi - p| \le |\lambda^q - 1| \le 2\pi |q\xi - p|$.) It follows that $(11 : 1)$ is equivalent to the requirement that

$$|\lambda^q - 1| > \epsilon'/q^{\kappa-1} \qquad \Longleftrightarrow \qquad 1/|\lambda^q - 1| < c\,q^{\kappa-1}$$

for some $\epsilon' > 0$, with the same value of κ and with $c = 1/\epsilon'$. Thus Siegel's Theorem 11.4 can be restated as follows:

> *If the angle $\xi \in \mathbb{R}/\mathbb{Z}$ is Diophantine of any order, then any holomorphic germ with multiplier $\lambda = e^{2\pi i \xi}$ is locally linearizable.*

Note that $\mathcal{D}(\kappa) \subset \mathcal{D}(\eta)$ whenever $\kappa < \eta$. It turns out that the $\mathcal{D}(\kappa) = \emptyset$ for $\kappa < 2$. (Compare Problem 11-a.) Diophantine numbers of order two are said to be *of bounded type*. (Compare 11.9 below.) Examples are provided by quadratic irrationals. More generally, we have the following classical statement.

11.6. Theorem of Liouville. *If the irrational number ξ satisfies a degree d polynomial equation $f(\xi) = 0$ with integer coefficients, then ξ is Diophantine of order d.*

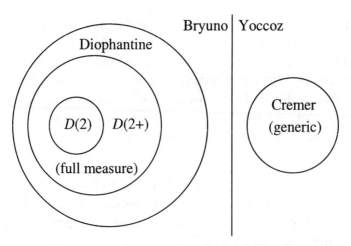

Figure 23. Schematic diagram for classes of irrational numbers.

Proof. We may assume that $f(p/q) \neq 0$. Clearing denominators, it follows that $|f(p/q)| \geq 1/q^d$. On the other hand, if M is an upper bound for $|f'(x)|$ in the interval of length one centered at ξ, then

$$|f(p/q)| \leq M |\xi - p/q| .$$

Choosing $\epsilon < 1/M$, we obtain $|\xi - p/q| > \epsilon/q^d$, as required. $\square$

Thus every algebraic number is Diophantine. It follows that any irrationally indifferent fixed point with algebraic rotation number is locally linearizable. (Compare Figure 22a.)

Remark. Irrational numbers which are not Diophantine are often called *Liouville numbers*. We will see in Appendix C that the set of Liouville numbers has Hausdorff dimension zero.

A much sharper version of 11.6 was proved by Klaus Roth in 1952. He showed that: *Every algebraic number belongs to the class $\mathcal{D}(2+) = \bigcap_{\kappa > 2} \mathcal{D}(\kappa)$, consisting of numbers which are Diophantine of order κ for every $\kappa > 2$.* We will need the following much more elementary property.

11.7. Lemma. *The set $\mathcal{D}(2+)$ has full measure on the circle $\mathbb{R}/\mathbb{Z}$.*

Evidently 11.4 and 11.7 together imply 11.5.

Proof of 11.7. Let $U(\kappa, \epsilon)$ be the open set consisting of all $\xi \in [0,1]$ such that $|\xi - p/q| < \epsilon/q^\kappa$ for some p/q. This set has measure at most

$$\sum_{q=1}^{\infty} q \cdot 2\epsilon/q^\kappa ,$$

since for each q there are q possible choices of p/q modulo one. If $\kappa > 2$, then this sum converges, and hence tends to zero as $\epsilon \searrow 0$. Therefore the intersection $\bigcap_{\epsilon > 0} U(\kappa, \epsilon)$ has measure zero, and its complement $\mathcal{D}(\kappa)$ has full measure. Taking the intersection of these complements $\mathcal{D}(\kappa)$ as $\kappa \searrow 2$, we see that the set $\mathcal{D}(2+)$ also has full measure. $\square$

On the other hand, we will see in Appendix C that the subset $\mathcal{D}(2)$ has measure zero.

For a more precise analysis of the approximation of an irrational number $\xi \in (0, 1)$ by rationals, we consider the *continued fraction expansion*

$$\xi = \cfrac{1}{a_1 + \cfrac{1}{a_2 + \cfrac{1}{a_3 + \cdots}}}$$

where the a_i are uniquely defined strictly positive integers. The rational number

$$\frac{p_n}{q_n} = \cfrac{1}{a_1 + \cfrac{1}{a_2 + \cfrac{\ddots}{\quad + \cfrac{1}{a_{n-1}}}}}$$

is called the n-th *convergent* to ξ. The denominators q_n will play a particularly important role. Here is a summary of results which will be proved in the Appendix. (Compare Hardy and Wright.)

11.8. Continued Fraction Theorem. *Each rational number p_n/q_n is the best possible approximation to ξ by fractions with denominator at most q_n. In fact we have the more precise statement that*

$$|\lambda^h - 1| > |\lambda^{q_n} - 1| \qquad for \qquad 0 < h < q_{n+1}, \quad h \neq q_n .$$

The error $|\lambda^{q_n} - 1|$ has the order of magnitude of $1/q_{n+1}$; in fact:

$$\frac{2}{q_{n+1}} \leq |\lambda^{q_n} - 1| \leq \frac{2\pi}{q_{n+1}} .$$

The denominators q_n can be computed inductively by the formula

$$q_{n+1} = a_n q_n + q_{n-1} \geq 2q_{n-1}, \quad with \quad q_0 = 0, \ q_1 = 1, \ q_2 = a_1 .$$

For proofs, see Appendix C.

11.9. Corollary. *An irrational number ξ is Diophantine if and only if q_{n+1} is less than some polynomial function of q_n. More precisely, it belongs to $\mathcal{D}(\kappa)$ if and only if q_{n+1} is less than some constant times $q_n^{\kappa-1}$. In particular, it belongs to $\mathcal{D}(2)$ if and only if the ratios q_{n+1}/q_n are bounded, or equivalently if and only if the continued fraction coefficients $a_n = (q_{n+1} - q_{n-1})/q_n$, are bounded.*

For this reason, elements of $\mathcal{D}(2)$ are also called *numbers of bounded type* (or sometimes "numbers of constant type"). The proof of 11.9 is straightforward, and will be left to the reader. $\square$

Next we state three results which give a fairly precise picture of the local problem. In 1972 Bryuno proved a much sharper version of 11.4.

11.10. Theorem of Bryuno. *If*

$$\sum_n \frac{\log(q_{n+1})}{q_n} < \infty, \qquad (11:2)$$

then any holomorphic germ with a fixed point of multiplier λ is locally linearizable.

Yoccoz, in 1987 showed that this is a best possible result, by completely analyzing the quadratic polynomial case.

11.11. Theorem of Yoccoz. *Conversely, if the sum $(11:2)$ diverges, then the quadratic map $f(z) = z^2 + \lambda z$ has a fixed point at the origin which is not locally linearizable. Furthermore, this fixed point has the "small cycles property": Every neighborhood of the origin contains infinitely many periodic orbits.*

Without attempting to prove these theorems, we can give some intuitive idea as to what they mean in the polynomial case as follows. Whenever the summand $(\log q_{n+1})/q_n$ is large the rotation number will be extremely close to p_n/q_n, so that f will be extremely close to a parabolic map with a period q_n cycle of repelling directions. It follows that the basin of infinity for f will have a period q_n cycle of deep fjords which penetrate towards zero, squeezing the size of a possible Siegel disk. As an example, in Figure 24, the summands $\log(31)/3 = 1.144\cdots$ and $\log(6200003)/31 = 0.504\cdots$ correspond to fjords of period 3 and 31 which are visible in the figure. When the sum $(11:2)$ is infinite, such a Siegel disk can no longer exist.

Historical Remark. One early contributer to considerations of this kind was T. M. Cherry, but only part of his work had been published at the time of his death in 1966. According to [Love]: "Fuller details of this

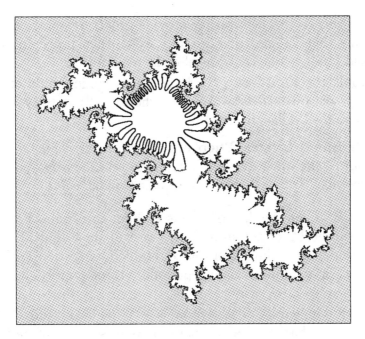

Figure 24. Quadratic Siegel disk with rotation number

$$1/(3 + 1/(10 + 1/(200000 + 1/\cdots))) .$$

The boundary of the Siegel disk has been emphasized. Note the fjords with period $q_2 = 3$ and $q_3 = 31$ which squeeze this disk.

may possibly be written in his notebooks; it is likely that he studied this subject deeply over many years". It is to be hoped that these notebooks will someday be made public.

Yoccoz's theorem raises the question as to whether every Cremer point has small cycles. The answer was provided by Perez-Marco in 1990. Suppose that $\sum \log(q_{n+1})/q_n = \infty$ so that a Cremer point can exist.

11.12. Theorem of Perez-Marco. *If*

$$\sum \frac{\log \log(q_{n+1})}{q_n} < \infty , \tag{11 : 3}$$

then every germ of a holomorphic function which has a Cremer point at the origin has this small cycles property. But in the generic case where the sum (11 : 3) diverges, there exists a germ with multiplier λ such that every forward orbit contained in some neighborhood of zero has zero as accumulation point. Evidently such a germ has no small cycles but is not linearizable.

We will not try to say any more about these three big theorems. The remainder of this section will rather provide proofs for some easier results. We first show that Cremer points really exist, then prove Cremer's 1927 theorem, and finally show that Siegel disks really exist.

First a folk theorem. Let $\{f_\lambda\}$ be a holomorphic family of non-linear rational maps parametrized by $\lambda \in \mathbb{C}$, where $f_\lambda(0) = 0$, $f_\lambda'(0) = \lambda$, so that $f_\lambda(z) = \lambda z + \text{(higher terms)}$.

11.13. Small Cycles Theorem. *For generic choice of λ on the unit circle, there are infinitely many periodic orbits in every neighborhood of $z = 0$, hence zero is a Cremer point.*

Proof. Let $\mathbb{D}_\epsilon$ be the disk of radius ϵ about the origin. For λ in some dense open subset U_ϵ of the unit circle, we will show that $\mathbb{D}_\epsilon$ contains a non-zero periodic orbit. For λ in the countable intersection $\bigcap U_{1/n}$, it will follow that f_λ has infinitely many periodic orbits converging to zero, as required.

Start with some root of unity $\lambda_0 = e^{2\pi i p/q} \neq 1$, and choose some positive $\epsilon' \leq \epsilon$ so that f_{λ_0} has no periodic points $z \neq 0$ of period $\leq q$ in the closed disk $\overline{\mathbb{D}}_{\epsilon'}$. Then the algebraic number of fixed points of $f_{\lambda_0}^{\circ k}$ in $\mathbb{D}_{\epsilon'}$ is equal to one for $1 \leq k < q$ (compare Problem 10-b), but is strictly greater than one for $k = q$. As we vary λ throughout a neighborhood of λ_0, this multiple fixed point for $f_{\lambda_0}^{\circ q}$ at the origin, will split up into a collection of fixed points for $f_\lambda^{\circ q}$. Let $U(p/q)$ be a neighborhood of λ_0 which is small enough so that no periodic point of period $\leq q$ can cross through the boundary of $D_{\epsilon'}$. Then for any $\lambda \in U(p/q)$, $\lambda \neq \lambda_0$ it follows that f_λ has an entire periodic orbit of period q contained within the neighborhood $\mathbb{D}_{\epsilon'} \subset \mathbb{D}_\epsilon$. The union U_ϵ of these open sets $U(p/q)$, for $0 < p/q < 1$, is evidently a dense and open subset of the circle, with the required property. $\square$

Proof of Cremer's Theorem 11.2. First consider a monic polynomial $f(z) = z^d + \cdots + \lambda z$ of degree $d \geq 2$ with a fixed point of multiplier λ at the origin. Then $f^{\circ q}(z) = z^{d^q} + \cdots + \lambda^q z$, so the fixed points of $f^{\circ q}$ are the roots of the equation

$$z^{d^q} + \cdots + (\lambda^q - 1)z = 0 .$$

Therefore, the product of the $d^q - 1$ non-zero fixed points of $f^{\circ q}$ is equal to $\pm(\lambda^q - 1)$. Choosing q so that $|\lambda^q - 1| < 1$, it follows that at least one of these fixed points z satisfies

$$0 < |z|^{d^q} < |z|^{d^q - 1} \leq |\lambda^q - 1| .$$

Therefore, if the quantity $\liminf |\lambda^q - 1|^{1/d^q}$ is zero, it follows that there exist periodic points $z \neq 0$ in every neighborhood of zero.

In order to extend this argument to the case of a rational function f of degree $d \geq 2$, Cremer first noted that f must map at least one point $z_1 \neq 0$ to the fixed point $z = 0$. After conjugating by a Möbius transformation which carries z_1 to infinity, we may assume that $f(\infty) = f(0) = 0$. If we set $f(z) = P(z)/Q(z)$, this means that P is a polynomial of degree strictly less than d. After a scale change, we may assume that $P(z)$ and $Q(z)$ have the form

$$P(z) = \star z^{d-1} + \cdots + \star z^2 + \lambda z, \qquad Q(z) = z^d + \cdots + 1,$$

where each $\star$ stands for a possibly non-zero coefficient. A brief computation then shows that $f^{\circ q}(z) = P_q(z)/Q_q(z)$ where $P_q(z)$ and $Q_q(z)$ have the form

$$P_q(z) = \star z^{d^q - 1} + \cdots + \star z^2 + \lambda^q z, \qquad Q_q(z) = z^{d^q} + \cdots + 1.$$

Thus the equation for fixed points of $f^{\circ q}$ has the form

$$0 = zQ_q(z) - P_q(z) = z(z^{d^q} + \cdots + (1 - \lambda^q)).$$

Now, if $\liminf |\lambda^q - 1|^{1/d^q} = 0$, then just as in the polynomial case we see that f has infinitely many periodic points in every neighborhood of zero, and hence that $0 \in J(f)$. This proves 11.2. $\square$

For the proof that 11.2 implies 11.3, see Problem 11-b.

Remark. As far as I know, Cremer never studied the small cycles property. His argument finds periodic points in every neighborhood of zero, but does not show that the entire periodic orbit is contained in a small neighborhood of zero. However, compare Problem 11-d.

Finally, let us show that Siegel disks really exist. We will describe a proof, due to Yoccoz, of the following special case of Siegel's Theorem. (Compare [Herman, 1986] or [Douady, 1987].) Let $f_\lambda(z) = z^2 + \lambda z$.

11.14. Theorem. *For Lebesgue almost every angle $\xi \in \mathbb{R}/\mathbb{Z}$, taking $\lambda = e^{2\pi i \xi}$ as usual, the quadratic map $f_\lambda(z)$ possesses a Siegel disk about the origin.*

The proof will depend on approximating multipliers λ on the unit circle by multipliers with $|\lambda| < 1$.

Definition. For each λ in the closed unit disk $\overline{\mathbb{D}}$, define the *size* $\sigma(\lambda)$ to be the largest number σ such that there exists a univalent map $\psi_\lambda : \mathbb{D}_\sigma \to \mathbb{C} \smallsetminus J(f_\lambda)$ from the open disk of radius σ into the Fatou set of

f_λ satisfying the following form of the Schröder equation,

$$f_\lambda(\psi_\lambda(w)) = \psi_\lambda(\lambda w) \qquad \text{for all} \quad w \in \mathbb{D}_\sigma,$$
$$\text{with} \qquad \psi_\lambda(0) = 0, \quad \psi_\lambda'(0) = 1, \qquad (11:4)$$

taking $\sigma(\lambda) = 0$ if such a map cannot exist for any positive radius. Evidently, $\sigma(\lambda) > 0$ whenever f_λ has a Siegel disk about the origin, and this number does indeed measure the size of the disk in some invariant sense. Similarly $\sigma(\lambda) > 0$ for $0 < |\lambda| < 1$. However, $\sigma(\lambda) = 0$ whenever f_λ has a parabolic or Cremer point at the origin, and also when $\lambda = 0$. If f_λ has a Siegel disk, note that this size function cannot be continuous at λ, since parabolic or Cremer values for λ are everywhere dense on the unit circle.

Recall that a real valued function σ on a topological space is said to be *upper semicontinuous* if

$$\limsup_{x \to x_0,\ x \neq x_0} \sigma(x) \leq \sigma(x_0)$$

for every x_0 in the space, or equivalently if the set of x with $\sigma(x) \geq \sigma_0$ is closed for every $\sigma_0 \in \mathbb{R}$.

11.15. Lemma. *This size function $\sigma : \overline{\mathbb{D}} \to \mathbb{R}$ is bounded and upper semicontinuous. Furthermore, for $|\lambda| < 1$ we can write $\sigma(\lambda) = |\eta(\lambda)|$, where the function $\lambda \mapsto \eta(\lambda)$ is holomorphic throughout the open unit disk.*

Proof. First note that $\sigma(\lambda) \leq 2$ for all $\lambda \in \overline{\mathbb{D}}$. In fact, if $|z| > 2$ and $|\lambda| \leq 1$, then $|f_\lambda(z)| = |z(z + \lambda)| > |z|$, and it follow easily that z lies in the basin of infinity for f_λ. Therefore any map $\mathbb{D}_\sigma \to \mathbb{C}$ satisfying (11 : 4) must take values in $\mathbb{D}_2$, and hence must satisfy $\sigma \leq 2$ by the Schwarz Lemma.

To see that σ is upper semicontinuous, note that $\sigma(z) \geq \sigma_0$ if and only if there is a univalent map $\mathbb{D}_{\sigma_0} \to \mathbb{D}_2$ satisfying (11 : 4). But the collection of all holomorphic maps from $\mathbb{D}_{\sigma_0}$ to $\mathbb{D}_2$ forms a normal family. Hence any sequence of such maps contains a subsequence which is locally uniformly convergent throughout $\mathbb{D}_{\sigma_0}$. In particular, given a sequence of univalent maps ψ_{λ_k} satisfying (11 : 4), we can find a convergent subsequence, and it is not hard to check that the limit function will also be univalent and satisfy (11 : 4). Therefore, the set of $\lambda \in \overline{\mathbb{D}}$ with $\sigma(\lambda) \geq \sigma_0$ is closed, as required.

Now let us specialize to the case $0 < |\lambda| < 1$. We can compute the size $\sigma(\lambda)$ for such values of λ as follows. Let $\mathcal{A}_\lambda$ be the attractive basin of the fixed point zero under f_λ. Then, as in 8.2 and 8.4, the Kœnigs map

$\phi_\lambda : \mathcal{A}_\lambda \to \mathbb{C}$ can be defined by the formula

$$\phi_\lambda(z) = \lim_{n \to \infty} f_\lambda^{\circ n}(z)/\lambda^n . \qquad (11:5)$$

Since this limit converges locally uniformly, $\phi_\lambda(z)$ depends holomorphically on both variables. (Compare 8.3.) In particular, its value

$$\eta(\lambda) = \phi_\lambda(-\lambda/2)$$

at the critical point of f_λ is a holomorphic function of λ. It follows easily from 8.5 that the absolute value $|\eta(\lambda)|$ is precisely equal to the size $\sigma(\lambda)$ as defined above for all $\lambda \in \mathbb{D} \smallsetminus \{0\}$. Furthermore, since $\sigma(0) = 0$, it follows from upper semicontinuity that $\eta(\lambda) \to 0$ as $\lambda \to 0$, so that η has a removable singularity at the origin. This completes the proof of 11.15. $\square$

Remark. We can actually compute this function within the open disk by noting that $\eta(\lambda)$ can be described as the limit as $i \to \infty$ of the numbers

$$\eta_i = f_\lambda^{\circ i}(-\lambda/2)/\lambda^i .$$

These can be determined recursively by the formula

$$\eta_0 = -\lambda/2 , \quad \eta_{i+1} = \eta_i + \lambda^{i-1}\eta_i^2 .$$

This procedure can also be used to compute the coefficients of the power series expansion of η about the origin, which takes the form

$$\eta(\lambda) = -\frac{\lambda}{4} + \frac{\lambda^2}{16} + \frac{\lambda^3}{16} + \frac{\lambda^4}{32} + \frac{9\lambda^5}{256} + \frac{\lambda^6}{256} + \frac{7\lambda^7}{256} - \frac{3\lambda^8}{512} + \cdots .$$

Details will be left to the reader.

11.16. Corollary. *For $|\lambda_0| = 1$, the map f_{λ_0} has either a Cremer point or a parabolic point at the origin if and only if the limit*

$$\lim_{\lambda \to \lambda_0 , \ |\lambda|<1} \eta(\lambda)$$

is defined and equal to zero.

The proof is immediate. Now, to complete the proof of 11.14, we need only quote the following classical result. Let c_0 be any complex constant.

Theorem of F. and M. Riesz. *If $\eta : \mathbb{D} \to \mathbb{C}$ is a non-constant bounded holomorphic function, then the set of angles $\xi \in \mathbb{R}/\mathbb{Z}$ such that the **radial limit***

$$\lim_{r \nearrow 1} \eta(r \exp(2\pi i \xi))$$

is defined and equal to c_0 has Lebesgue measure zero.

For the proof of this result, see A.3 in Appendix A. Clearly Theorem 11.14 follows immediately. □

Remark. This argument shows also that

$$\sigma(e^{2\pi i \xi}) \geq \limsup_{r \nearrow 1} |\eta(re^{2\pi i \xi})|$$

for every $\xi \in \mathbb{R}/\mathbb{Z}$. In fact Yoccoz has shown that equality always holds here.

Unsolved Problems: Although the theorems of Bryuno, Yoccoz and Perez-Marco are very sharp, they do not answer all questions about local behavior near an irrationally indifferent fixed point. For example, there is a very complicated local structure about any Cremer point (compare [Perez-Marco, 1997]), yet there is not a single example which is well understood. It is not known whether any rational function can have a Cremer point without small cycles. Also, it is not known whether any non-linear rational function can have a Siegel disk for which the Bryuno condition is not satisfied.

In 8.6 and 10.11, we found a rather direct relationship between critical points and attracting or parabolic orbits. For Siegel or Cremer orbits, the relationship is less direct.

Definition. By the *postcritical set* $P = P(f)$ of a rational map f we will mean the union of all forward images $f^{\circ k}(c)$ with $k > 0$, where c ranges over the critical points. We will be particularly interested in the topological closure $\overline{P}(f)$ of this set.

> **11.17. Theorem.** *Every Cremer fixed point or periodic point for a rational map is contained in the postcritical closure $\overline{P}(f)$. Similarly, the boundary of any Siegel disk or cycle of Siegel disks is contained in $\overline{P}(f)$.*

Proof. (Compare §19.) We will work with the open sets $U = \hat{\mathbb{C}} \smallsetminus \overline{P}$ and $V = f^{-1}(U)$. Since $f^{-1}(\overline{P}) \supset \overline{P}$, it follows that $V \subset U$. Since there are no critical values in U, it follows that f maps V onto U by an d-fold covering map; or more precisely that f maps each connected component of V onto some connected component of U by a covering map.

We may assume that $\overline{P}$ contains at least three distinct points. For otherwise U would be a twice punctured sphere, hence its covering space V would also be a twice punctured sphere, equal to U. It would then follow easily that f must be conjugate to a map of the form $z \mapsto z^{\pm d}$, with no Cremer points and no Siegel disks.

Thus we may assume that every connected component of V or U is conformally hyperbolic. Consider a fixed point $z_0 = f(z_0)$ which belongs to

U and hence to V. Let $V_0 \subset U_0$ be the connected components containing z_0. If $V_0 = U_0$, then V_0 maps into itself under f, and hence is contained in the Fatou set. Thus in this case z_0 cannot be a Cremer point.

Now suppose that V_0 is strictly smaller than U_0. Then the inclusion $V_0 \to U_0$ strictly decreases Poincaré distances by $(2:6)$, that is

$$\mathrm{dist}_V(x, y) > \mathrm{dist}_U(x, y)$$

for every $x \neq y$ in V_0. On the other hand, by 2.11, V maps to U by a local isometry, so that

$$\mathrm{dist}_U(f(x), f(y)) = \mathrm{dist}_V(x, y)$$

whenever $x, y \in V$ are sufficiently close to each other. Therefore

$$\mathrm{dist}_U(f(x), f(y)) > \mathrm{dist}_U(x, y) \qquad (11:6)$$

whenever $x \neq y$ in V_0 are sufficently close to each other. It follows that the fixed point z_0 must be strictly repelling: Again it cannot be a Cremer point.

To deal with the boundary of a Siegel disk $\Delta = f(\Delta)$, we must work just a little harder. First note that Δ, with its center point z_0 removed, is naturally foliated into f-invariant circles. The intersection $\overline{P} \cap \Delta \smallsetminus \{z_0\}$ consists at most of finitely many of these circles. Thus if a component U_0 of $\hat{\mathbb{C}} \smallsetminus \overline{P}$ intersects the boundary $\partial\Delta$, then it must contain an entire neighborhood of $\partial\Delta$ within Δ. In particular, it must contain every invariant circle C which is sufficiently close to the boundary. Similarly, one component V_0 of $f^{-1}(U_0)$ must contain every such circle. If $V_0 = U_0$ then, arguing as above, U_0 is contained in the Fatou set and cannot intersect the boundary of Δ. On the other hand, if V_0 is strictly smaller than U_0 then, as in $(11:6)$, f restricted to V_0 must strictly increase the distance $\mathrm{dist}_U(x, y)$ between nearby points, and similarly must map any smooth path to a path of strictly larger arclength. In particular, it must map each invariant circle $C \subset V_0$ onto a longer circle. But this is impossible since f maps C diffeomorphically onto itself. This proves that every fixed Cremer point or Siegel disk boundary must be contained in $\overline{P}$. The corresponding statement for a cycle of Cremer points or Siegel disks follows by applying the above argument to a suitable iterate $f^{\circ k}$, and noting that $P(f^{\circ k}) = P(f)$. $\square$

A different proof of 11.17 will be given in 14.4.

_____

Here are some problems for the reader.

Problem 11-a (Dirichlet). Use the "pigeon-hole principle" to show that for any irrational number x there are infinitely many fractions p/q with

$$\left| x - \frac{p}{q} \right| < \frac{1}{q^2} .$$

In fact, for any integer $Q > 1$ cut the circle $\mathbb{R}/\mathbb{Z}$ into Q half-open intervals of length $1/Q$, and consider the $Q+1$ numbers 0, x, $2x$, $\ldots$, Qx reduced modulo $\mathbb{Z}$. Since at least two of these must belong to the same interval, conclude that there exist integers p and $1 \leq q \leq Q$ with $|qx - p| < 1/Q$, hence

$$\left| x - \frac{p}{q} \right| < \frac{1}{qQ} \leq \frac{1}{q^2} .$$

Problem 11-b. Generic Angles. Given a completely arbitrary sequence of positive real numbers ϵ_1, ϵ_2, $\ldots$ $\searrow 0$, let $S(q_0)$ be the set of all real numbers ξ such that

$$\left| \xi - \frac{p}{q} \right| < \epsilon_q$$

for some fraction p/q in lowest terms with $q > q_0$. Show that $S(q_0)$ is a dense open subset of $\mathbb{R}$, and conclude that the intersection $S = \bigcap_{q_0} S(q_0)$, consisting of all ξ for which this condition is satisfied for infinitely many p/q, is a countable intersection of dense open sets. As an example, taking $\epsilon_q = 2^{-q!}$ conclude that a generic real number belongs to the set S, and hence satisfies Cremer's condition that $\liminf |\lambda^q - 1|^{1/d^q} = 0$ for every degree d. (Compare 11.2.)

Problem 11-c. Cremer's 1938 Theorem. If $f(z) = \lambda z + a_2 z^2 + a_3 z^3 + \cdots$, where λ is not zero and not a root of unity, show (following Poincaré) that there is one and only one formal power series of the form $h(z) = z + h_2 z^2 + h_3 z^3 + \cdots$ which formally satisfies the condition that $h(\lambda z) = f(h(z))$. In fact

$$h_n = \frac{a_n + X_n}{\lambda^n - \lambda}$$

for $n \geq 2$, where $X_n = X(a_2, \ldots, a_{n-1}, h_2, \ldots, h_{n-1})$ is a certain polynomial expression whose value can be computed inductively. Now suppose that we choose the a_n inductively, always equal to zero or one, so that $|a_n + X_n| \geq 1/2$. If

$$\liminf_{q \to \infty} |\lambda^q - 1|^{1/q} = 0 ,$$

show that the uniquely defined power series $h(z)$ has radius of convergence zero. Conclude that $f(z)$ is a holomorphic germ which is not locally linearizable. Choosing the a_n more carefully, show that we can even choose $f(z)$ to be an entire function.

Problem 11-d. Small Cycles. Suppose that

$$\limsup_{q\to\infty} \frac{\log\log(1/|\lambda^q - 1|)}{q} > \log d > 0 .$$

Modify the proof of 11.2 to show that: *Any fixed point of multiplier λ for a rational function f of degree d has the small cycle property.* First choose $\epsilon > 0$ so that

$$\log\log(1/|\lambda^q - 1|) > (\epsilon + \log d)q ,$$

or equivalently

$$|\lambda^q - 1|^{1/d^q} < \exp(-e^{\epsilon q}) ,$$

for infinitely many q. The proof of 11.2 then constructs points z_q of period q with $|z_q| < \exp(-e^{\epsilon q})$. Now use Taylor's Theorem to find $\delta > 0$ so that $|f(z)| < e^\epsilon |z|$ for $|z| < \delta$, hence $|f^{\circ q}(z)| < \delta$ for $|z| < e^{-q\epsilon}\delta$. Finally, note that $\exp(-e^{\epsilon q}) < e^{-q\epsilon}\delta$ for large q, to conclude that f has small cycles.

PERIODIC POINTS: GLOBAL THEORY

§12. The Holomorphic Fixed Point Formula

First recall the following. Let $f : \widehat{\mathbb{C}} \to \widehat{\mathbb{C}}$ be a rational map of degree $d \geq 0$.

12.1. Lemma. *If f is not the identity map, then f has exactly $d+1$ fixed points, counted with multiplicity.*

Here the *multiplicity* of a finite fixed point $z_0 = f(z_0)$ is defined to be the unique integer $m \geq 1$ for which the power series expansion of $f(z) - z$ about z_0 has the form

$$f(z) - z = a_m(z - z_0)^m + a_{m+1}(z - z_0)^{m+1} + \cdots$$

with $a_m \neq 0$. Thus $m \geq 2$ if and only if the multiplier λ at z_0 is exactly 1. (Note that z_0 is then a parabolic point with $m-1$ attracting petals, each of which maps into itself. Compare §10.) In the special case of a fixed point at infinity, we introduce the local uniformizing parameter $\zeta = \phi(z) = 1/z$, and define the multiplicity of f at infinity to be the multiplicity of the map $\phi \circ f \circ \phi^{-1}$ at the point $\phi(\infty) = 0$. As an example, any polynomial map of degree $d \geq 2$ has a fixed point at infinity with multiplier $\lambda = 0$ and hence with multiplicity $m = 1$, and therefore has d finite fixed points counted with multiplicity. On the other hand, the map $f(z) = z + 1$ has a fixed point of multiplicity $m = 2$ at infinity.

Proof of 12.1. Conjugating f by a fractional linear automorphism if necessary, we may assume that the point at infinity is *not* fixed by f. If we write f as a quotient $f(z) = p(z)/q(z)$ of two polynomials which have no common factor, this means that the degrees of $p(z)$ and $q(z)$ satisfy

$$\text{degree}(p(z)) \leq \text{degree}(q(z)) = d.$$

Evidently the equation $f(z) = z$ is equivalent to the polynomial equation $p(z) = z \, q(z)$ of degree $d+1$, hence it has $d+1$ solutions, counted with multiplicity. $\square$

Remark. This algebraic multiplicity m is known to topologists as the *Lefschetz fixed point index*. For any map $f : M \to M$ of a compact n-dimensional manifold into itself with only finitely many fixed points, the Lefschetz indices of these fixed points are defined, and their sum is equal to

$$\sum_{i=0}^{n} (-1)^i \, \text{trace}\Big(f_* : H_i(M; \mathbb{R}) \to H_i(M; \mathbb{R})\Big).$$

(See for example Franks.) In our case, with M the Riemann sphere and f

rational of degree d, there is a contribution of $+1$ from the 0-dimensional homology and $+d$ from the 2-dimensional homology, so the sum of the indices is $d + 1$.

Both Fatou and Julia made use of a "well known" relation between the multipliers at the fixed points of a rational map. First consider an isolated fixed point $z_0 = f(z_0)$ where $f : U \to \mathbb{C}$ is a holomorphic function on a connected open set $U \subset \mathbb{C}$. The *residue fixed point index* of f at z_0 is defined to be the complex number

$$\iota(f, z_0) = \frac{1}{2\pi i} \oint \frac{dz}{z - f(z)} \qquad (12:1)$$

where we integrate in a small loop in the positive direction around z_0.

Lemma 12.2. *If the multiplier $\lambda = f'(z_0)$ is not equal to $+1$, then this residue fixed point index is given by*

$$\iota(f, z_0) = \frac{1}{1 - \lambda}. \qquad (12:2)$$

Proof. Without loss of generality, we may assume that $z_0 = 0$. Expanding f as a power series, we can write

$$f(z) = \lambda z + a_2 z^2 + a_3 z^3 + \cdots,$$

hence

$$\frac{z - f(z)}{z} = (1 - \lambda) - a_2 z - a_3 z^2 - \cdots,$$

and

$$\frac{z}{z - f(z)} = \frac{1}{1 - \lambda} + b_1 z + b_2 z^2 + \cdots$$

for suitable coefficients b_j, or in other words

$$\frac{1}{z - f(z)} = \frac{1}{(1 - \lambda)z} + b_1 + b_2 z + \cdots.$$

Integrating this expression around a small loop $|z| = \epsilon$, we evidently obtain a residue of $1/(1 - \lambda)$, as asserted. $\square$

Note: This computation breaks down completely in the special case $\lambda = 1$. The index $\iota(f, z_0)$ is still well defined and finite, but the formula $(12:2)$ no longer makes sense. Compare Problems 12-a, 12-b.

More generally, given any isolated fixed point of a holomorphic map $F : S \to S$ from a Riemann surface to itself, we can choose some local coordinate z, and then compute the index $\iota(f, z_0)$ for the associated local map $z \mapsto f(z)$.

Lemma 12.3. *This residue, computed in terms of a local co-ordinate near the fixed point, does not depend on any particular choice of local coordinate.*

In the generic case of a fixed point with multiplier $\lambda \neq 1$, this follows immediately from 12.2, since the multiplier clearly does not depend on the particular choice of coordinate chart. A proof which works also when $\lambda = 1$ will be given at the end of this section.

Now suppose that our Riemann surface S is the Riemann sphere. (For analogous formulas on other Riemann surfaces, see 12.5.)

12.4. Rational Fixed Point Theorem. *For any rational $f : \hat{\mathbb{C}} \to \hat{\mathbb{C}}$ which is not the identity map, we have the relation*

$$\sum_{z=f(z)} \iota(f, z) = 1,$$

to be summed over all fixed points.

Here is an easy application, to illustrate this Fixed Point Formula: Suppose that f has one fixed point with multiplier very close to 1, and hence with $|\iota|$ large. Then f must have at least one other fixed point with $|\iota|$ large, and hence with λ close to 1 or equal to 1.

Proof of 12.4. Conjugating f by a linear fractional automorphism if necessary, we may assume that $f(\infty) \neq 0, \infty$. Then $f(z)$ converges to $f(\infty) \in \mathbb{C} \smallsetminus \{0\}$ as $z \to \infty$, hence

$$\frac{1}{z - f(z)} - \frac{1}{z} = \frac{f(z)}{z(z - f(z))} \sim \frac{f(\infty)}{z^2}$$

as $z \to \infty$. Let $L(r)$ be the loop $|z| = r$. It follows easily that the integral of this difference around $L(r)$ converges to zero as $r \nearrow \infty$. Hence

$$\frac{1}{2\pi i} \oint_{L(r)} \frac{dz}{z - f(z)} = \frac{1}{2\pi i} \oint_{L(r)} \frac{dz}{z} = +1,$$

Evidently the integral on the left is equal to the sum of the residues $\iota(f, z_j)$ at the various fixed points of f; hence the required summation formula.

Examples. A rational map $f(z) = c$ of degree zero has just one fixed point, with multiplier zero and hence with index $\iota(f, c) = 1$. A rational map of degree one usually has two distinct fixed points, and the relation

$$\frac{1}{1 - \lambda_1} + \frac{1}{1 - \lambda_2} = 1$$

simplifies to $\lambda_1 \lambda_2 = 1$. Thus such a map can have at most one attracting fixed point. Any polynomial map $p(z)$ of degree two or more has a superattracting fixed point at infinity, with multiplier zero and hence with index $\iota(p, \infty) = 1$. *Thus the sum of the indices at the finite fixed points for a non-linear polynomial map is always zero.*

In the case of a polynomial of degree two, the relation

$$\frac{1}{1 - \lambda_1} + \frac{1}{1 - \lambda_2} = 0$$

for the finite fixed points simplifies to $\lambda_1 + \lambda_2 = 2$. As an example, for the family of quadratic maps

$$f_\lambda(z) = z^2 + \lambda z$$

the multipliers at the two fixed points are λ at $z = 0$, and $2 - \lambda$ at $z = 1 - \lambda$. *Thus this map f_λ has an attracting fixed point if and only if λ belong either to the unit disk $\mathbb{D}$ centered at $\lambda = 0$, or to the disk $2 + \mathbb{D}$ centered at $\lambda = +2$.* These two disks are clearly visible in Figure 25, which shows the λ-parameter plane. (Compare Appendix F.)

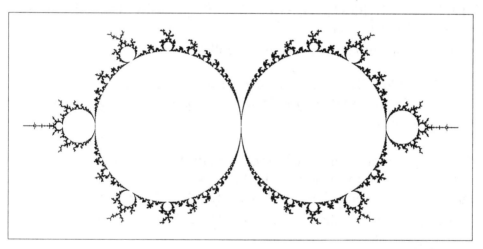

Figure 25. "Double Mandelbrot set": the bifurcation locus in the λ-parameter plane for the family of quadratic maps $z \mapsto z^2 + \lambda z$. The figure is centered at $\lambda = 1$.

Remark 12.5. There is a far-reaching generalization of this fixed point theorem, due to Atiyah and Bott. In particular, for a holomorphic map f from a compact Riemann surface of genus g to itself, the Atiyah-Bott formula implies that the sum of the residue fixed point indices is given by

$$\sum \iota = 1 - \overline{\tau}, \qquad\qquad (12:3)$$

where τ is the trace of the induced map from the g-dimensional vector space of holomorphic 1-forms to itself, and where the overline stands for complex conjugation. In the special case of the Riemann sphere, with $g = 0$, there are no holomorphic 1-forms, so this formula reduces to 12.4. For other examples, see Problem 12-d.

12.6. Lemma. *A fixed point with multiplier $\lambda \neq 1$ is attracting if and only if its residue fixed point index ι has real part $\mathrm{Re}(\iota) > \frac{1}{2}$.*

Geometrically, this is proved by noting that a fixed point with multiplier λ is attracting if and only if $1 - \lambda$ belongs to the disk $1 + \mathbb{D}$, having the origin as boundary point. It is easy to check that the map $z \mapsto 1/z$ carries this disk $1 + \mathbb{D}$ precisely onto the half-plane $\mathrm{Re}(z) > 1/2$. Computationally, it can be proved by noting that $\frac{1}{2} < \mathrm{Re}(\frac{1}{1-\lambda})$ if and only if

$$ 1 < \frac{1}{1-\lambda} + \frac{1}{1-\bar{\lambda}} . $$

Multiplying both sides by $(1-\lambda)(1-\bar{\lambda}) > 0$, we easily obtain the equivalent inequality $\lambda\bar{\lambda} < 1$. $\square$

One important consequence is the following.

12.7. Corollary. *Every rational map of degree $d \geq 2$ must have either a repelling fixed point, or a parabolic fixed point with $\lambda = 1$, or both.*

Proof. If there is no fixed point of multiplier $\lambda = 1$, then there must be $d + 1$ distinct fixed points. If these were all attracting or indifferent, then each index would have real part $\mathrm{Re}(\iota) \geq \frac{1}{2}$, hence the sum would have real part greater than or equal to $\frac{d+1}{2} > 1$; but this would contradict 12.2. $\square$

Since repelling points and parabolic points both belong to the Julia set, this yields a constructive proof of the following. (Compare 4.5.)

12.8. Corollary. *The Julia set for a non-linear rational map is always non-vacuous.*

To conclude this section, we must prove 12.3. That is, we must prove that the residue $\iota(f, z_0)$ is invariant under a local holomorphic change of coordinate $w = \phi(z)$. In fact we will give both a computational proof and a geometric proof.

Computational Proof of 12.3. Set $\eta(z) = z - f(z)$ and use the power series expansion $\phi(f(z)) = \phi(z - \eta) = \phi(z) - a_1\eta + a_2\eta^2 - + \cdots$, where $a_k = a_k(z)$ is $1/k!$ times the k-th derivative of ϕ at z. Setting

$F(w) = F(\phi(z)) = \phi(f(z))$, we have

$$\iota(F, \phi(z_0)) = \frac{1}{2\pi i} \oint \frac{dw}{w - F(w)} = \frac{1}{2\pi i} \oint \frac{\phi'(z)\, dz}{\phi(z) - \phi(f(z))}$$

$$= \frac{1}{2\pi i} \oint \frac{a_1\, dz}{a_1\, \eta - a_2\, \eta^2 + - \cdots}$$

$$= \frac{1}{2\pi i} \oint \left(1 + \frac{a_2}{a_1}\eta + \frac{a_2^2 - a_1 a_3}{a_1^2}\eta^2 + \cdots\right)\frac{dz}{\eta} \; ,$$

with $a_1 \neq 0$, where both η and the a_j are holomorphic functions of z . Evidently this last integral is equal to

$$\frac{1}{2\pi i} \oint \frac{dz}{\eta} = \iota(f, z_0) \; . \quad \square$$

Geometric Proof. Taking $z_0 = 0$ for convenience, suppose that we can find a one-parameter family of maps $f_t : \overline{\mathbb{D}}_\epsilon \to \mathbb{C}$ so that f coincides with f_0 on the disk $\overline{\mathbb{D}}_\epsilon$, and so that f_t has only simple fixed points for $t \neq 0$. (Thus the fixed point of multiplicity $m \geq 2$ for f_0 must split up into m simple fixed points for small values of $t \neq 0$.) We may assume that ϵ is small enough so that f_0 has only the single fixed point in $\overline{\mathbb{D}}_\epsilon$. If $|t|$ is sufficiently small, then f_t will have no fixed points on the boundary circle $\partial \mathbb{D}_\epsilon$, and the sum of the fixed point indices for f_t in $\mathbb{D}_\epsilon$ can be expressed as an integral around the boundary which depends continuously on the parameter t . For $t \neq 0$ this sum has the form $\sum 1/(1 - \lambda_j)$ where each λ_j is a holomorphic conjugacy invariant. It follows that

$$\iota(f_0, 0) = \lim_{t \to 0} \sum_1^m 1/(1 - \lambda_j)$$

is also a holomorphic conjugacy invariant.

To construct an example of such a one-parameter family, we can simply set $f_t(z) = f(z) - t$. Since the derivative $f'(z)$ is not identically equal to $+1$, we can choose ϵ so that $f'(z) \neq 1$ for $0 < |z| \leq \epsilon$, and it follows that $f_t : \overline{\mathbb{D}}_\epsilon \to \mathbb{C}$ has only simple fixed points for $t \neq 0$. $\square$

Problem 12-a. If $f(z) = z + \alpha z^2 + \beta z^3 + \text{(higher terms)}$, with $\alpha \neq 0$, show that the residue index is given by $\iota(f, 0) = \beta/\alpha^2$. As an example, consider the one-parameter family of cubic maps

$$f_\alpha(z) = z^3 + \alpha z^2 + z$$

with a double fixed point at the origin. Using 12.4 or by direct calculation, show that the remaining finite fixed point $z = -\alpha$ has multiplier $\lambda = 1 + \alpha^2$, and hence is attracting if and only if α^2 lies within a unit disk centered at -1, or if and only if α lies within a figure eight shaped region bounded by a lemniscate. This lemniscate is clearly visible as the boundary of the main upper and lower regions in Figure 26, which shows the α-parameter plane. Now, for $\alpha \neq 0$, suppose that we perturb f_α to a map $z \mapsto z^3 + \alpha z^2 + (1 - \epsilon) z$, so that the double fixed point at the origin splits up into two distinct nearby fixed points. First suppose that α^2 lies inside the disk of radius $1/2$ centered at $1/2$, or equivalently that α lies within a corresponding region bounded by a lemniscate shaped like the symbol ∞. (This has been drawn in as a dotted line in Figure 26.) Show that we can choose a small $\epsilon \in \mathbb{C}$ so that *both* of the fixed points near zero are attracting. On the other hand, if α lies strictly outside this region, show for any such perturbation that at least one of the fixed points near zero must be repelling. (Compare the description of "parabolic attracting" or "repelling" points in [A. Epstein].)

Figure 26. Parameter space picture for the family of cubic maps $z \mapsto z^3 + \alpha z^2 + z$. *In the outer region, the orbit of one critical point escapes to infinity.*

Problem 12-b. More generally, consider a fixed point of multiplicity $n + 1 \geq 2$ which has been put into the normal form

$$f(z) = z + \alpha z^{n+1} + \beta z^{2n+1} + \text{(higher terms)} .$$

(Compare Problem 10-d. After a scale change, we may assume that $\alpha = 1$.) Show that the index $\iota(f, 0)$ is equal to the ratio β/α^2.

Problem 12-c. Any fixed point z_0 for f is evidently also a fixed point for $f^{\circ k}$. If z_0 is attracting [or repelling], show that $\iota(f^{\circ k}, z_0)$ tends to the limit 1 [or 0] as $k \to \infty$. For a fixed point of multiplicity $m \geq 2$, show that $\iota(f^{\circ k}, z_0)$ tends to the limit $m/2$.

Problem 12-d. Verify the generalized fixed point formula of 12.5 in the following two special cases:

If $f : \mathbb{T} \to \mathbb{T}$ is a linear torus map with derivative f' identically equal to α, show that the trace τ of the induced action on the 1-dimensional space of holomorphic 1-forms is equal to α. If f is not the identity map, show that there are $|1 - \alpha|^2$ fixed points, each with index $\iota = 1/(1 - \alpha)$, and conclude that $\sum \iota = 1 - \overline{\tau}$, as required. (Compare Problem 6-b.)

Now suppose that S is a compact surface of genus g and that $f : S \to S$ is an involution with k fixed points. Use the Riemann-Hurwitz formula to conclude that the quotient S/f is a surface of genus $\hat{g} = (2 + 2g - k)/4$. For the induced action on the g-dimensional vector space of holomorphic 1-forms, show that $\hat{g}$ of the eigenvalues are equal to $+1$ that the remaining $g - \hat{g}$ are equal to -1, so that the trace τ equals $2\hat{g} - g$. Conclude that $\sum \iota = k/2 = 1 - \overline{\tau}$.

§13. Most Periodic Orbits Repel

This section will prove the following theorem of Fatou. By a *cycle* we will mean simply a periodic orbit of f. Recall that a cycle is called *attracting*, *indifferent*, or *repelling* according as its multiplier λ satisfies $|\lambda| < 1$, $|\lambda| = 1$, or $|\lambda| > 1$.

13.1. Theorem. *Let $f : \widehat{\mathbb{C}} \to \widehat{\mathbb{C}}$ be a rational map of degree $d \geq 2$. Then f has at most a finite number of cycles which are attracting or indifferent.*

We will see in §14 that there always exist infinitely many repelling cycles. Shishikura has given the sharp upper bound of $2d - 2$ for the number of attracting or indifferent cycles, using methods of quasi-conformal surgery. (Compare A. Epstein.) However the classical proof, which is given here, shows only that this number is less than or equal to $6d - 6$.

First recall from 10.12 that f can have at most $2d - 2$ attracting or parabolic cycles. (If $\mathcal{A}$ is any immediate attracting or parabolic basin, then some iterate $f^{\circ p}$ maps $\mathcal{A}$ into itself. By 8.6 or 10.11, $\mathcal{A}$ contains a critical point of $f^{\circ p}$. Therefore, by the chain rule, some immediate basin $f^{\circ i}(\mathcal{A})$ in the same cycle contains a critical point of f. Since the various immediate basins are all pairwise disjoint, and since f has at most $2d - 2$ critical points, it follows that f has at most $2d - 2$ cycles which are attracting or parabolic.)

13.2. Lemma. *For a rational map of degree $d \geq 2$, the number of indifferent cycles which have multiplier $\lambda \neq 1$ is at most $4d - 4$.*

Evidently 13.2 and 10.12 together imply Theorem 13.1. Following Fatou, we prove 13.2 by perturbing the given map f in such a way that more than half of its indifferent cycles become attracting. Let $f(z) = p(z)/q(z)$ where $p(z)$ and $q(z)$ are polynomials, at least one of which has degree d. Consider the one-parameter family of maps

$$f_t(z) = \frac{p(z) + tz^d}{q(z) + t}, \qquad (13:1)$$

with $f_0(z) = f(z)$. For most values of the parameter t this is a well defined rational map of degree d which depends smoothly on t, tending to the limit $f_\infty(z) = z^d$ as $t \to \infty$. However, we must exclude a finite number of exceptional parameter values for which a zero and a pole of f_t crash together: First, we must exclude any t for which the numerator and denominator of $(13:1)$ can vanish simultaneously. But if the denominator is zero then $t = -q(z)$, and if the numerator is also zero then

$$p(z) - q(z)z^d = 0 .$$

If we exclude the trivial case where $f(z)$ is identically equal to z^d, then this last equation has only finitely many roots z_j, hence we must exclude the finitely many parameter values $t = -q(z_j)$. This is enough to control behavior at finite values of z. If $q(z)$ has degree less than d, then to control behavior near $z = \infty$ we must also exclude the unique parameter value t_0 such that $p(z) + t_0 z^d$ has degree less than d.

If $f = f_0$ has k distinct indifferent cycles with multipliers $\lambda_j \neq 1$, then we must prove that $k \leq 4d - 4$. Choose a representative point z_j in each of these cycles. By the Implicit Function Theorem, we can follow each of these cycles under a small deformation of f_0. Thus, for small values of $|t|$, the map f_t must have corresponding periodic points $z_j(t)$ with multipliers $\lambda_j(t)$ which depend holomorphically on t, with $|\lambda_j(0)| = 1$.

13.3. Sub-Lemma. *None of these functions $t \mapsto \lambda_j(t)$ can be constant throughout a neighborhood of $t = 0$.*

Proof. Suppose that for some j the function $t \mapsto \lambda_j(t)$ were constant throughout a neighborhood of $t = 0$. Choose some ray $r \mapsto t = re^{i\theta}$ from 0 to ∞, where $0 \leq r \leq \infty$, which avoids the finitely many exceptional values of t. Then we will show that it is possible to continue the function $t \mapsto z_j(t)$ analytically along a neighborhood of this ray, so that $z_j(t)$ is a periodic point for f_t with multiplier $\lambda_j = \text{constant}$. To prove this, we will check that the set of $r_1 \in [0, \infty]$ such that we can continue for $0 \leq r \leq r_1$ is both open and closed: It is closed since any limit point of periodic points with fixed multiplier $\lambda_j \neq 1$ is itself a periodic point with this same multiplier, and it is open since any such periodic point varies smoothly with t throughout some open neighborhood in the t-plane by the Implicit Function Theorem. Now continuing analytically along the ray to $t = \infty$, we see that the map $z \mapsto z^d$ must also have a cycle with multiplier equal to λ_j, with $|\lambda_j| = 1$. But every periodic point of this limit map is either 0 or ∞ with multiplier $\lambda = 0$, or else a root of unity with $\lambda = d^k > 1$. This contradiction completes the proof of 13.3. $\square$

The proof of 13.2 continues as follows. We can express each of our k multipliers as a locally convergent power series

$$\lambda_j(t)/\lambda_j(0) = 1 + a_j t^{n_j} + (\text{higher terms}) ,$$

where $a_j \neq 0$ and $n_j \geq 1$. Hence

$$|\lambda_j(t)| = 1 + \text{Re}(a_j t^{n_j}) + o(t^{n_j}) .$$

We can divide the t-plane up into n_j sectors for which the expression $\operatorname{Re}(a_j t^{n_j})$ is positive, and n_j complementary sectors for which this expression is negative. Let

$$\sigma_j(\theta) \;=\; \operatorname{sgn}\!\left(\operatorname{Re}(a_j e^{i\theta n_j})\right),$$

so that

$$\sigma_j(\theta) \;=\; +1 \qquad \Longrightarrow \qquad |\lambda_j(re^{i\theta})| > 1 \quad \text{for small } r > 0,$$
$$\sigma_j(\theta) \;=\; -1 \qquad \Longrightarrow \qquad |\lambda_j(re^{i\theta})| < 1 \quad \text{for small } r > 0.$$

Evidently each $\sigma_j : \mathbb{R}/2\pi\mathbb{Z} \to \{\pm 1, 0\}$ is a step function which takes the value ± 1 except at $2n_j$ jump discontinuities, with average

$$\frac{1}{2\pi} \int_0^{2\pi} \sigma_j(\theta)\, d\theta \;=\; 0.$$

Therefore the sum $\sigma_1(\theta) + \cdots + \sigma_k(\theta)$ is also a well defined step function with average zero. Replacing k by $k-1$ if necessary, we may assume that k is odd, so that this sum takes odd values almost everywhere. Hence we can choose some θ such that $\sigma_j(\theta) = -1$ for more than half, that is for at least $(k+1)/2$, of the indices j. If we choose r sufficiently small and set $t = re^{i\theta}$, this means that f_t has at least $(k+1)/2$ distinct cycles with multiplier satisfying $|\lambda_j| < 1$. Therefore, by 8.6 or 10.12, we have $(k+1)/2 \le 2d-2$, or in other words $k+1 \le 4d-4$. This completes the proof of 13.2 and 13.1. $\square$

§14. Repelling Cycles are Dense in J

We saw in 4.3 that every repelling cycle is contained in the Julia set. The following much sharper statement was proved by Fatou and by Julia.

14.1. Theorem. *The Julia set for any rational map of degree ≥ 2 is equal to the closure of its set of repelling periodic points.*

Since the proofs by Julia and by Fatou are interesting and different, we will give both.

Proof following Julia. We will use the Rational Fixed Point Formula of §12. Recall from 12.7 that every rational map f of degree two or more has either a repelling fixed point, or a fixed point with $\lambda = 1$. In either case, this fixed point belongs to the Julia set $J(f)$. (Compare 4.3, 4.4.)

Thus we can start with a fixed point z_0 in the Julia set. Let $U \subset \hat{\mathbb{C}}$ be any open set, disjoint from z_0, which intersects $J(f)$. The next step is to construct a special orbit $\cdots \mapsto z_2 \mapsto z_1 \mapsto z_0$ which passes through U and terminates at this fixed point z_0. By definition, such an orbit is called *homoclinic* if the backwards limit $\lim_{j \to \infty} z_j$ exists and is equal to the terminal point z_0. To construct a homoclinic orbit, we will appeal to Theorem 4.10 which says that there exists an integer $r > 0$ and a point $z_r \in J(f) \cap U$ so that the r-th forward image $f^{\circ r}(z_r)$ is equal to z_0. Given any neighborhood N_0 of z_0, we can repeat this argument and conclude that there exists an integer $q > r$ and a point $z_q \in N_0$ so that $f^{\circ(q-r)}(z_q) = z_r$. (Figure 27.)

To be more explicit, in the case where z_0 is a repelling fixed point we choose N_0 to be a linearizing neighborhood, as in the Kœnigs Theorem 8.2. In the parabolic case, we choose N_0 to be a flower neighborhood, as in 10.5. In either case, we choose N_0 small enough to be disjoint from z_r. It then follows that we can inductively choose preimages $\cdots \mapsto z_j \mapsto z_{j-1} \mapsto \cdots \mapsto z_q$, all inside of the neighborhood N_0. *These preimages z_j will automatically converge to z_0 as $j \to \infty$.* If z_0 is repelling, this is clear. In the parabolic case, z_q cannot belong to an attracting petal; hence it must belong to a repelling petal, and again this statement is clear.

First suppose that none of the points $\cdots \mapsto z_j \mapsto \cdots \mapsto z_0$ in this homoclinic orbit are critical points of f. Then a sufficiently small disk neighborhood V_q of $z_q \in N_0$ will map diffeomorphically under $f^{\circ q}$ onto a neighborhood V_0 of z_0. Pulling this neighborhood V_q back under iterates of f^{-1}, we obtain neighborhoods $z_j \in V_j$ for all j, shrinking down towards the limit point z_0 as $j \to \infty$. In particular, if we choose p sufficiently large, then $\bar{V}_p \subset V_0$. Now f^{-p} maps the simply connected

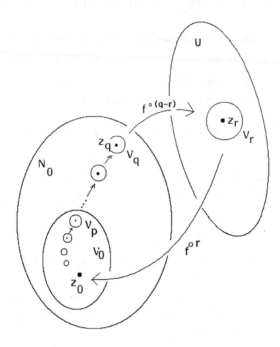

Figure 27. A homoclinic orbit.

open set V_0 holomorphically into a compact subset of itself. Hence it contracts the Poincaré metric on V_0 by a factor $c < 1$, and therefore must have an attractive fixed point z' within V_p. Evidently this point $z' \in V_p$ is a repelling periodic point of period p under the map f. Since the orbit of z' under f intersects the required open set U, the conclusion follows.

If our homoclinic orbit contains critical points, then this argument must be modified very slightly as follows. We can still choose simply connected neighborhoods V_j of the z_j so that $\bar{V}_p \subset V_0$ for some large p, and so that f maps each V_j onto V_{j-1}. However, some finite number of these mappings will be branched. Choose a slit S in V_0 from the boundary to the midpoint z_0 so as to be disjoint from $\bar{V}_p$, and choose some sector in V_p which maps isomorphically onto $V_0 - S$ under $f^{\circ p}$. The proof now proceeds just as before. $\square$

Proof of 14.1 following Fatou. In this case, the main idea is an easy application of Montel's Theorem 3.7. However, we must use Theorem 13.1 to finish the argument.

To begin the proof, recall from 4.11 that the Julia set $J(f)$ has no isolated points. Hence we can exclude finitely many points of $J(f)$ without affecting the argument. Let z_0 be any point of $J(f)$ which is not a fixed point, and not a critical value. In other words, we assume that there are d preimages $z_1, \ldots, z_d$, which are distinct from each other and from z_0, where $d \geq 2$ is the degree. By the Inverse Function Theorem, we can find d holomorphic functions $z \mapsto \varphi_j(z)$ which are defined throughout some neighborhood N of z_0, and which satisfy $f(\varphi_j(z)) = z$, with $\varphi_j(z_0) = z_j$. We claim that for some $n > 0$ and for some $z \in N$ the function $f^{on}(z)$ must take one of the three values z, $\varphi_1(z)$ or $\varphi_2(z)$. For otherwise the family of holomorphic functions

$$g_n(z) = \frac{(f^{on}(z) - \varphi_1(z))\,(z - \varphi_2(z))}{(f^{on}(z) - \varphi_2(z))\,(z - \varphi_1(z))}$$

on N would avoid the three values $0, 1$ and ∞, and hence be a normal family. (This expression is just the cross-ratio of the four points z, $\phi_1(z)$, $\phi_2(z)$, $f^{on}(z)$, as discussed in Problem 1-c.) It would then follow easily that $\{f^{on}|N\}$ was also a normal family, contradicting the hypothesis that N intersects the Julia set. Thus we can find $z \in N$ so as to satisfy either $f^{on}(z) = z$ or $f^{on}(z) = \varphi_j(z)$. Clearly it follows that z is a periodic point of period n or $n+1$ respectively.

This shows that every point in $J(f)$ can be approximated arbitrarily closely by periodic points. Since all but finitely many of these periodic points must repel, this completes the proof. $\square$

There are a number of interesting corollaries.

14.2. Corollary. *If U is an open set which intersects the Julia set J of f, then for n sufficiently large the image $f^{on}(U \cap J)$ is equal to the entire Julia set J.*

Proof. We know that U contains a repelling periodic point z_0 of period say p. Thus z_0 is fixed by the iterate $g = f^{op}$. Choose a small neighborhood $V \subset U$ of z_0 with the property that $V \subset g(V)$. Then clearly $V \subset g(V) \subset g^{o2}(V) \subset \cdots$. But it follows from 4.7 that the union of the open sets $g^{on}(V)$ contains the entire Julia set $J = J(f) = J(g)$. Since J is compact, this implies that $J \subset g^{on}(V) \subset g^{on}(U)$ for n sufficiently large, and the corresponding statement for f follows. $\square$

More generally, if $K \subset \hat{\mathbb{C}}$ is any compact set which does not contain any grand orbit finite points, then $f^{on}(U) \supset K$ for large n. In particular, if there are no grand orbit finite points, then $f^{on}(U) = \hat{\mathbb{C}}$ for large n. (Compare 4.7 and Problem 4-b.)

As another corollary, we can make a sharper statement of the defining property for the Julia set.

14.3. Corollary. *If $U \subset \widehat{\mathbb{C}}$ is any open set which intersects the Julia set $J(f)$, then no sequence of iterates $f^{\circ n(i)}$ can converge locally uniformly throughout U.*

Proof. Suppose that the sequence of functions $f^{\circ n(i)}(z)$ converged locally uniformly to $g(z)$ throughout the open set U. If z_0 is a point of $U \cap J$, then we could choose a smaller neighborhood U' of z_0 so that $|g(z) - g(z_0)| < \epsilon$ for all $z \in U'$. For large i, it would follow that $|f^{n(i)}(z) - g(z_0)| < 2\epsilon$ for all $z \in U'$, which is impossible by 14.2. $\square$

As another consequence of 14.2, we can give an alternative proof of Theorem 11.17 (with a slightly sharper statement in the Cremer case). By a *critical orbit* we will mean the forward orbit of some critical point.

14.4. Corollary. *Every boundary point of a Siegel disk Δ belongs to the closure of some critical orbit. Similarly, any Cremer point is a non-isolated point in the closure of some critical orbit.*

Proof. If $z_0 \in \partial\Delta$ were not in the closure of some critical orbit, then we could construct a small disk V around z_0 so that the forward orbits of all critical points avoid V. This would mean that every branch of the n-fold iterated inverse function f^{-n} could be defined as a single valued holomorphic function $f^{-n} : V \to \widehat{\mathbb{C}}$. Let us choose that particular branch which carries the intersection $\Delta \cap V$ into Δ, necessarily by a "rotation" of Δ. Since the rotation number is irrational, we can choose some subsequence of iterated inverse maps which converges to the identity map on $\Delta \cap V$. This is evidently a normal family, since it avoids the central part of Δ. Hence there is a sub-subsequence $\{f^{-n(i)}\}$ which converges locally uniformly on all of V, necessarily to the identity map of V. It follows easily that the corresponding sequence of forward iterates $f^{\circ n(i)}$ also converges to the identity on V. But this contradicts 14.2.

In the Cremer case, we proceed as follows. (This argument applies also to parabolic cycles. Compare 10.11.) If the assertion were false, we could choose a small disk V around the given point z_0 so that no critical orbit intersects the punctured disk $V \smallsetminus \{z_0\}$. Replacing f by some iterate if necessary, we may assume that z_0 is fixed by f. Arguing as above, there exists a unique holomorphic branch $f^{-n} : V \to \widehat{\mathbb{C}}$ of the n-fold iterated inverse function which fixes the point z_0. These inverse maps form a normal family since, for example, they avoid any periodic orbit which is disjoint from V. Thus we can choose a subsequence $f^{-n(i)}$ converging locally uniformly to some holomorphic map $h : V \to \widehat{\mathbb{C}}$. By the Inverse function

Theorem, since $|h'(z_0)| = 1$, this h must map some small neighborhood of z_0 isomorphically onto a neighborhood V' of z_0. It follows that the corresponding forward maps $f^{\circ n(i)}$ converge on V' to the inverse map $h^{-1} : V' \to V$. Again, this contradicts 14.2. □

By definition, the rational map f is called *postcritically finite* if it has the property that every critical orbit is finite, or in other words is either periodic or eventually periodic. According to Thurston, such a map can be uniquely specified by a finite topological description. (Compare Douady & Hubbard, 1993.)

> **14.5. Corollary.** *If f is postcritically finite, then every periodic orbit of f is either repelling or superattracting. More generally, suppose that f has the property that every critical orbit either is finite, or converges to an attracting periodic orbit. Then every periodic orbit of f is either repelling or attracting; there are no parabolic cycles, Cremer cycles, or Siegel cycles.*

Proof. This follows immediately from 10.11 and 14.4. □

As still another consequence, we get a simpler proof of 4.12.

> **14.6. Corollary.** *If a Julia set J is not connected, then it has uncountably many distinct connected components.*

Proof. Suppose that J is the union $J_0 \cup J_1$ of two disjoint non-vacuous compact subsets. After replacing f by some iterate $g = f^{\circ n}$, we may assume by 14.2 that $g(J_0) = J$ and $g(J_1) = J$. Now to each point $z \in J$ we can assign an infinite sequence of symbols

$$\epsilon_0(z),\ \epsilon_1(z),\ \epsilon_2(z),\ \ldots\ \in\ \{0,1\}$$

by setting $\epsilon_k(z)$ equal to zero or one according as $g^{\circ k}(z)$ belongs to J_0 or J_1. It is not difficult to check that points with different symbol sequences must belong to different connected components of J, and that all possible symbol sequences actually occur. □

14.7. Concluding Remark. The statement that the Julia set is equal to the closure of the set of repelling periodic points is actually true for an arbitrary holomorphic map of an arbitrary Riemann surface, providing that we exclude just one rather trivial exceptional case. For transcendental functions this was proved by Baker (1968), and for maps of a torus it follows easily from 6.1. The unique exceptional case occurs for a fractional linear transformation of $\hat{\mathbb{C}}$ which has just one parabolic fixed point — for example the map $f(z) = z + 1$ with $J(f) = \{\infty\}$.

STRUCTURE OF THE FATOU SET

§15. Herman Rings

The next two sections will be surveys only, with no proofs for several major statements. This section will describe a close relative of the Siegel disk.

Definition. A component U of the Fatou set $\hat{\mathbb{C}} \smallsetminus J(f)$ is called a *Herman ring* if U is conformally isomorphic to some annulus $\mathcal{A}_r = \{z : 1 < |z| < r\}$, and if f (or some iterate of f) corresponds to an irrational rotation of this annulus. (Siegel disks and Herman rings are often collectively called "*rotation domains*".)

There are two known methods for constructing Herman rings. The original method, due to Herman, is based on a careful analysis of real analytic diffeomorphisms of the circle. An alternative method, due to Shishikura, uses quasiconformal surgery, starting with two copies of the Riemann sphere with a Siegel disk in each, cutting out part of the center of each disk and pasting the resulting boundaries together in order to fabricate such a ring.

The original method can be outlined as follows. (Compare Herman 1979, Sullivan 1983, Douady 1987-88.) First a number of definitions. If $f : \mathbb{R}/\mathbb{Z} \to \mathbb{R}/\mathbb{Z}$ is an orientation preserving homeomorphism, then we can lift to a homeomorphism $F : \mathbb{R} \to \mathbb{R}$ which satisfies the identity $F(t+1) = F(t) + 1$, and is uniquely defined up to addition of an integer constant.

Definition. The real number

$$\mathrm{Rot}(F) \;=\; \lim_{n \to \infty} \frac{F^{\circ n}(t_0)}{n}$$

is independent of the choice of t_0, and will be called the *translation number* of the lifted map F. Following Poincaré, the *rotation number* $\mathrm{rot}(f) \in \mathbb{R}/\mathbb{Z}$ of the circle map f is defined to be the residue class of $\mathrm{Rot}(F)$ modulo $\mathbb{Z}$.

It is well known that this construction is well defined, and invariant under orientation preserving topological conjugacy, and that it has the following properties. (Compare Coddington and Levinson, or de Melo and van Strien, and see Problem 15-d.)

> **15.1. Lemma.** *The homeomorphism f has a periodic point with period q if and only if its rotation number is rational with denominator q.*

> **15.2. Denjoy's Theorem.** *If f is a diffeomorphism of class*

C^2 , and if the rotation number $\rho = \mathrm{rot}(f)$ is irrational, then f is topologically conjugate to the rotation $t \mapsto t + \rho \pmod{\mathbb{Z}}$.

15.3. Lemma. *Consider a one-parameter family of lifted maps of the form*

$$F_\alpha(t) = F_0(t) + \alpha .$$

Then the translation number $\mathrm{Rot}(F_\alpha)$ increases continuously and monotonically with α , increasing by $+1$ as α increases by $+1$. (However, this dependence is not strictly monotone. Rather, there is an interval of constancy corresponding to each rational value of $\mathrm{Rot}(F_\alpha)$ provided that F_0 is non-linear.)

In the real analytic case, Denjoy's Theorem has an analog which can be stated as follows. Recall from §11 that a real number ξ is said to be *Diophantine* if there exist a (large) number n and a (small) number ϵ so that the distance of ξ from every rational number p/q satisfies $|\xi - p/q| > \epsilon/q^n$. The following was proved in a local version (that is for maps close to the identity) by Arnold, and sharpened first by Herman and then by Yoccoz.

15.4. Herman-Yoccoz Theorem. *If f is a real analytic diffeomorphism of $\mathbb{R}/\mathbb{Z}$ and if the rotation number ρ is Diophantine, then f is real analytically conjugate to the rotation $t \mapsto t + \rho \pmod{1}$.*

I will not attempt to give a proof. (In the C^∞ case, Herman and Yoccoz prove a corresponding if and only if statement: *Every C^∞ diffeomorphism with rotation number ρ is C^∞-conjugate to a rotation if and only if ρ is Diophantine.*)

Next we will need the concept of a *Blaschke product*. (Compare Problem 7-b, as well as 1.7.) Given any constant $a \in \hat{\mathbb{C}}$ with $|a| \neq 1$, it is not difficult to show that there is one and only one fractional linear transformation $z \mapsto \beta_a(z)$ which maps the unit circle $\partial\mathbb{D}$ onto itself fixing the base point $z = 1$, and which maps a to $\beta_a(a) = 0$. For example $\beta_0(z) = z$, $\beta_\infty(z) = 1/z$, and in general

$$\beta_a(z) = \frac{1 - \bar{a}}{1 - a} \cdot \frac{z - a}{1 - \bar{a}z}$$

whenever $a \neq \infty$. If $|a| < 1$, then β_a preserves orientation on the circle, and maps the unit disk into itself. On the other hand, if $|a| > 1$, then β_a reverses orientation on $\partial\mathbb{D}$ and maps $\mathbb{D}$ to its complement.

15.5. Lemma. *A rational map of degree d carries the unit circle into itself if and only if it can be written as a "Blaschke* *product"*

$$f(z) = e^{2\pi i t} \beta_{a_1}(z) \cdots \beta_{a_d}(z) \qquad (15:1)$$

for some constants $e^{2\pi i t} \in \partial \mathbb{D}$ and $a_1, \ldots, a_d \in \hat{\mathbb{C}} \smallsetminus \partial \mathbb{D}$.

Here the a_i must satisfy the conditions that $a_j \bar{a}_k \neq 1$ for all j and k. For if $a \bar{b} = 1$, then a brief computation shows that $\beta_a(z) \beta_b(z) \equiv 1$. Evidently the expression in 15.5 is unique, since the constants $e^{2\pi i t} = f(1)$ and $\{a_1, \ldots, a_d\} = f^{-1}(0)$ are uniquely determined by f. The proof of 15.5 is not difficult: Given f, one simply chooses any solution to the equation $f(a) = 0$, then divides $f(z)$ by $\beta_a(z)$ to obtain a rational map of lower degree, and continues inductively. $\square$

Such a Blaschke product carries the unit disk into itself if and only if all of the a_j satisfy $|a_j| < 1$. (Compare Problems 7-b, 15-c.) However, we will rather be interested in the mixed case, where some of the a_j are inside the unit disk and some are outside.

15.6. Theorem. *For any odd degree $d \geq 3$ we can choose a Blaschke product f of degree d which carries the unit circle $\partial \mathbb{D}$ into itself by an orientation preserving diffeomorphism with any desired rotation number ρ. If this rotation number ρ is Diophantine, then f possesses a Herman ring.*

Proof Outline. Let $d = 2n + 1$, and choose the a_j so that $n + 1$ of them are close to zero while the remaining n are close to ∞. Then it is easy to check that the Blaschke product $z \mapsto \beta_{a_1}(z) \cdots \beta_{a_d}(z)$ is C^1-close to the identity map on the unit circle $\partial \mathbb{D}$. In particular, it induces an orientation preserving diffeomorphism of $\partial \mathbb{D}$. Now multiplying by $e^{2\pi i t}$ and using 15.3, we can adjust the rotation number to be any desired constant. If this rotation number ρ is Diophantine, then there is a real analytic diffeomorphism h of $\partial \mathbb{D}$ which conjugates f to the rotation $z \mapsto e^{2\pi i \rho} z$. Since h is real analytic, it extends to a complex analytic diffeomorphism on some small neighborhood of $\partial \mathbb{D}$, and the conclusion follows. $\square$

As an example, Figure 28 shows the Julia set for the cubic rational map $f(z) = e^{2\pi i t} z^2 (z - 4)/(1 - 4z)$ with zeros at $0, 0, 4$, where the constant $t = .6151732 \cdots$ is adjusted so that the rotation number will be equal to $(\sqrt{5} - 1)/2$. There is a critical point near the center of this picture, with a Herman ring to its left, surrounding the superattractive basin about the origin in the left center. This is the simplest kind of example one can find, since Shishikura has shown that such a ring can exist only if the degree d is at least three (compare [Milnor, 1999b]), and since it is easy to check that a polynomial map cannot have any Herman ring. (Problem 15-a.)

Figure 28. Julia set for a cubic rational map possessing a Herman ring.

The rings constructed in this way are very special in that they are symmetric about the unit circle, with $f(1/\bar{z}) = 1/\bar{f}(z)$. However Herman's original construction, based on work of Helson and Sarason, was more flexible. Shishikura's more general construction also avoids the need for symmetry. Furthermore Shishikura's construction makes it clear that the possible rotation numbers for Herman rings are exactly the same as the possible rotation numbers for Siegel disks. In particular, any number satisfying the Bryuno condition of 11.10 can occur. The idea is roughly that one starts with two rational maps having Siegel disks with rotation numbers $+\rho$ and $-\rho$ respectively. One cuts out a small concentric disk from each, and then glues the resulting boundaries together. After making corresponding modifications at each of the infinitely many iterated pre-images of each of the Siegel disks, Shishikura applies the Morrey-Ahlfors-Bers Measurable Riemann Mapping Theorem in order to conjugate the resulting topological picture to an actual rational map.

Although Herman rings do not contain any critical points, none-the-less they are closely associated with critical points.

15.7. Lemma. *If U is a Herman ring, then every boundary point of U belongs to the closure of the orbit of some critical point. The boundary ∂U has two connected components, each of which is an infinite set.*

The proof of the first statement is almost identical to the proof of 11.17 or 14.4, while the second follows from Problem 5-b and the Jordan curve theorem. □

Problem 15-a. Using the maximum modulus principle, show that no polynomial map can have a Herman ring.

Problem 15-b. For any Blaschke product $f : \widehat{\mathbb{C}} \to \widehat{\mathbb{C}}$ show that z is a critical point of f if and only if $1/\bar{z}$ is a critical point, and show that z is a zero of f if and only if $1/\bar{z}$ is a pole.

Problem 15-c. A holomorphic map $f : \mathbb{D} \to \mathbb{D}$ is said to be *proper* if the inverse image of any compact subset of $\mathbb{D}$ is compact. Show that any proper holomorphic map from $\mathbb{D}$ onto itself can be expressed uniquely as a Blaschke product $(15 : 1)$, with $a_j \in \mathbb{D}$.

Problem 15-d. Show that the rotation number $\mathrm{rot}(f)$ can be deduced directly from the cyclic order relations on a single orbit, in a form covenient for computer calculations, as follows. Choose representatives $t_i \in [0, 1)$ for the elements of the orbit of zero, so that $t_i \equiv f^{\circ i}(0) \pmod{\mathbb{Z}}$. If we exclude the trivial case $t_1 = 0$, then t_1 cuts $[0, 1)$ into two disjoint intervals $I_1 = [0, t_1)$ and $I_0 = [t_1, 1)$. Define a sequence of bits $(b_2, b_3, b_4, \ldots)$ by the requirement that $t_n \in I_{b_n}$. If F is the unique lift with $F(0) = t_1$, show that

$$\mathrm{Rot}(F) = \lim_{n \to \infty} (b_2 + b_3 + \cdots + b_n)/n .$$

Furthermore, if a second such map f' has bit sequence $(b'_2, b'_3, \ldots)$, and if

$$(b_2, b_3, \ldots) < (b'_2, b'_3, \ldots) ,$$

using the lexicographical order for bit sequences, show that $\mathrm{Rot}(F) \leq \mathrm{Rot}(F')$.

§16. The Sullivan Classification of Fatou Components

The results in this section are due in part to Fatou and Julia, but with very major contributions by Sullivan.

A *Fatou component* for a nonlinear rational map f will mean any connected component of the Fatou set $\widehat{\mathbb{C}} \smallsetminus J(f)$. Evidently f carries each Fatou component U onto some Fatou component U' by a proper holomorphic map. First consider the special case $U = U'$.

16.1. Theorem. *If f maps the Fatou component U onto itself, then there are just four possibilities, as follows: Either U is the immediate basin for an attracting fixed point, or for one petal of a parabolic fixed point which has multiplier $\lambda = 1$, or else U is a Siegel disk or Herman ring.*

Here we are lumping together the case of a superattracting fixed point, with multiplier $\lambda = 0$, and the case of a geometrically attracting fixed point, with $\lambda \neq 0$. Note that immediate attractive or parabolic basins always contain critical points by 6.6 and 7.11, while rotation domains (that is Siegel disks and Herman rings) evidently cannot contain critical points.

Much of the proof of 16.1 has already been carried out in §5. In fact, according to 5.2 and 5.5, a priori there are just four possibilities. Either:

(a) U contains an attractive fixed point;

(b) all orbits in U converge to a boundary fixed point;

(c) f is an automorphism of finite order; or

(d) f is conjugate to an irrational rotation of a disk, punctured disk, or annulus.

In Case (a) we are done. Case (c) cannot occur, since our standing hypothesis that the degree is two or more guarantees that there are only countably many periodic points. In Case (d) we cannot have a punctured disk, since the puncture point would have to be a fixed point belonging to the Fatou set, so that U would be a subset of a Siegel disk, rather than a full Fatou component. Thus, in order to prove 16.1, we need only show that the boundary fixed point in Case (b) must be parabolic with $\lambda = 1$. This boundary fixed point certainly cannot be an attracting point or a Siegel point, since it belongs to the Julia set. Furthermore, it cannot be repelling, since it attracts all orbits in U. Thus it must be indifferent, $|\lambda| = 1$. To prove 16.1, we need only show that λ is precisely equal to $+1$.

The proof will be based on the following statement, which is due to Douady and Sullivan. (Compare Sullivan 1983, or Douady-Hubbard 1984-

5, p. 70. For a more classical alternative, see Lyubich 1986, p. 72.) Let

$$f(z) = \lambda z + a_2 z^2 + a_3 z^3 + \cdots$$

be a map which is defined and holomorphic in some neighborhood V of the origin, and which has a fixed point with multiplier λ at $z = 0$. By a *path in $V \smallsetminus \{0\}$ which converges to the origin* we will mean a continuous map $p : [0, \infty) \to V \smallsetminus \{0\}$ satisfying the condition that $p(t)$ tends to zero as $t \to \infty$. (Here $[0, \infty)$ denotes the half-open interval consisting of all real numbers $t \geq 0$.) Note that such a path p may have self-intersections.

16.2.   Snail Lemma.   *Suppose that there exists a path $p : [0, \infty) \to V \smallsetminus \{0\}$ which is mapped into itself by f in such a way that $f(p(t)) = p(t+1)$, and which converges to the origin as $t \to \infty$. Then either $|\lambda| < 1$ or $\lambda = 1$.*

In other words, the origin must be either an attracting fixed point, or a parabolic fixed point with λ precisely equal to 1.

Proof of 16.2. By hypothesis, the orbit $p(0) \mapsto p(1) \mapsto p(2) \mapsto \cdots$ in $V \smallsetminus \{0\}$ converges towards the origin. Thus the origin certainly cannot be a repelling fixed point; we must have $|\lambda| \leq 1$. Let us assume that $|\lambda| = 1$ with $\lambda \neq 1$, and show that this hypothesis leads to a contradiction.

As the path $t \mapsto p(t)$ winds closer and closer to the origin, the behavior of the map f on $p(t)$ is more and more dominated by the linear term $z \mapsto \lambda z$. Thus we have the asymptotic equality $p(t+1) \sim \lambda p(t)$ as $t \to \infty$. If the path p has no self-intersections, then the image must resemble a very tight spiral as shown in Figure 29(left), and we can sketch a proof as follows. Draw a radial segment E joining two turns of this spiral, as shown. Then the region W bounded by E together with a segment of the spiral will be mapped strictly into itself by f. Therefore, by the Schwarz Lemma, the fixed point of f at the point $0 \in W$ must be strictly attracting; which contradicts the hypothesis that $|\lambda| = 1$.

In order to fill in the details of this argument, and to allow for the possibility of self-intersections, let us introduce polar coordinates (r, θ) on $\mathbb{C} \smallsetminus \{0\}$, setting $z = re^{i\theta}$ with $r > 0$ and $\theta \in \mathbb{R}/2\pi\mathbb{Z}$. Lift p to a path $\tilde{p}(t) = (r(t), \tilde{\theta}(t))$ in the universal covering of $\mathbb{C} \smallsetminus \{0\}$, where $\tilde{\theta}(t)$ is a real number. As t tends to infinity, note that

$$r(t) \searrow 0 \qquad \text{and} \qquad \tilde{\theta}(t+1) = \tilde{\theta}(t) + c + o(1), \qquad (6:1)$$

where c is a uniquely defined real constant with $e^{ic} = \lambda$. Similarly, choosing $r_0 > 0$ so that f is univalent on the disk of radius r_0, we can lift f to a map $\tilde{f}(r, \tilde{\theta}) = (r', \tilde{\theta}')$ on the universal covering, where $\tilde{f}$ is

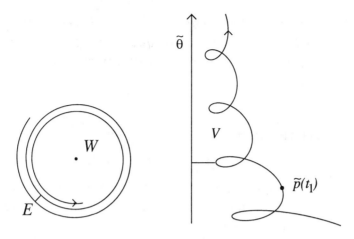

Figure 29. Simple curve in $\mathbb{C} \smallsetminus \{0\}$ on the left, and a non-simple curve lifted to the universal covering on the right.

defined and univalent for $0 < r < r_0$ and for all $\tilde{\theta} \in \mathbb{R}$. Note that

$$r' \sim r \qquad \text{and} \qquad \theta' = \theta + c + o(1) \qquad \text{as} \qquad r \searrow 0 \,,$$

where c is the same constant which occurs in $(6 : 1)$ provided that we choose the correct lift of f. (It follows that we can extend $\tilde{f}$ continuously over $[0, r_0) \times \mathbb{R}$ so that it is a translation, $\tilde{f}(0, \tilde{\theta}) = (0, \tilde{\theta} + c)$, when $r = 0$.)

We must prove that $c = 0$. Suppose for example that c were strictly positive. Then we could derive a contradiction as follows. Choose a constant $r_1 < r_0$ so that the map $\tilde{f}(r, \tilde{\theta}) = (r', \tilde{\theta}')$ satisfies $\tilde{\theta}' > \tilde{\theta} + c/2$ whenever $r \leq r_1$. Then choose t_1 so that $r(t) \leq r_1$ and hence $\tilde{\theta}(t+1) > \tilde{\theta}(t) + c/2$ whenever $t \geq t_1$. Furthermore, choose $\tilde{\theta}_1$ so that $\tilde{\theta}(t) < \tilde{\theta}_1$ for $t \leq t_1$. Now in the $(r, \tilde{\theta})$ plane, consider the connected region V which lies above the line $\tilde{\theta} = \tilde{\theta}_1$, to the right of the line $r = 0$, and to the left of the curve $\tilde{p}[t_1, \infty)$. It follows easily that $\tilde{f}$ maps V univalently into itself, and that the r-coordinate tends to zero under iteration of $\tilde{f}$ restricted to V. Hence, if W is the image of V under projection to the z-plane, it follows that W is a neighborhood of the origin, and that all orbits in W converge to the origin. Therefore the origin is an attracting fixed point, and $|\lambda| < 1$. $\square$

Here is a completely equivalent statement, which will be useful in §18. Again let f be a holomorphic map of the form $f(z) = \lambda z + a_2 z^2 + \cdots$ near the origin.

16.3. Corollary. *Now suppose that* $p : [0, \infty) \to V \smallsetminus \{0\}$ *is a path which converges to the origin, with* $f(p(t)) = p(t - 1)$ *for* $t \geq 1$ *(so that points on this path are pushed away from the origin). Then the multiplier* λ *must satisfy either* $|\lambda| > 1$ *or* $\lambda = 1$.

Proof. Since the orbit

$$\cdots \mapsto p(2) \mapsto p(1) \mapsto p(0)$$

is repelled by the origin, the multiplier λ cannot be zero. Hence f^{-1} is defined and holomorphic near the origin. Applying 16.2 to the map $g = f^{-1}$, the conclusion follows. $\square$

Proof of 16.1. Recall that we have already discussed all of the cases except (b) above. Thus we need only consider a Fatou component U which is mapped into itself by f in such a way that all orbits converge to a boundary fixed point w_0 . Choose any base point z_0 in U , and choose any path $p : [0, 1] \to U$ from $z_0 = p(0)$ to $f(z_0) = p(1)$. Extending for all $t \geq 0$ by setting $p(t+1) = f(p(t))$, we obtain a path in U which converges to the boundary point w_0 as $t \to \infty$. Therefore, according to 16.2, the fixed point w_0 must be either parabolic with $\lambda = 1$, or attracting. But w_0 belongs to the Julia set, and hence cannot be attracting. $\square$

Thus we have classified the Fatou components which are mapped onto themselves by f . There is a completely analogous description of Fatou components which cycle periodically under f . These are just the Fatou components which are fixed by some iterate of f . Each one is either:

(1) the immediate attractive basin for some attracting periodic point,

(2) the immediate basin for some petal of a parabolic periodic point,

(3) one member of a cycle of Siegel disks, or

(4) one member of a cycle of Herman rings.

In Cases (3) and (4), the topological type of the domain U is uniquely specified by this description. In Cases (1) and (2) as noted in 8.7 and Problem 10-e, U must be either simply connected or infinitely connected.

By §10, there can be at most a finite number of attracting basins and Siegel disks. Sullivan showed also that there can be at most a finite number of Herman rings, and hence that there are altogether only finitely many periodic Fatou components. (More precisely, according to Shishikura, there can be at most 2d-2 distinct cycles of periodic Fatou components.)

In order to complete the picture, we need the following fundamental theorem, which asserts that there are no *"wandering"* Fatou components.

16.4. Sullivan Non-Wandering Theorem. *Every Fatou component U for a rational map is eventually periodic. That is, there necessarily exist integers $n \geq 0$ and $p \geq 1$ so that the n-th forward image $f^{\circ n}(U)$ is mapped onto itself by $f^{\circ p}$.*

Thus every Fatou component is a preimage, under some iterate of f, of a component of one of the four types described above. The proof, by quasiconformal deformation, will be outlined in Appendix F. (Compare Sullivan 1985, Carleson and Gamelin. The intuitive idea can be outlined briefly as follows: If a wandering Fatou component were to exist, then using the Measurable Riemann Mapping Theorem of Morrey, Ahlfors and Bers one could construct an infinite dimensional space of deformations, all of which would have to be rational maps of the same degree. But the space of rational maps of fixed degree is finite dimensional.)

Recall from 14.5 that f is *postcritically finite* if every critical orbit is finite.

16.5. Corollary. *If a postcritically finite rational map has no superattractive periodic orbit, then its Julia set is the entire sphere $\widehat{\mathbb{C}}$.*

For by 8.6 and 10.11 it cannot have any attracting or parabolic basins, and by 11.17 and 15.7 it cannot have any rotation domains. □

We will give a more direct proof for this statement in 19.8. (See also Problem 16-e.)

Remark on Transcendental Maps. The analogues of both 16.1 and 16.4 fail for the iterates of a transcendental map $f : \mathbb{C} \to \mathbb{C}$. In fact there are two new kinds of Fatou component, which cannot occur for rational maps. There may be wandering domains (Problem 16-c), and there may be invariant domains $U = f(U)$ such that no orbit in U has any accumulation point in the finite plane $\mathbb{C}$. (Problem 16-d. These are now known as *Baker domains*. Of course every orbit in U must have an accumulation point in $\widehat{\mathbb{C}}$, but this point at infinity is an essential singularity of f, and hence looks very different from a parabolic point.)

Problem 16-a. Limits of iterates. Give a sharper formulation of the defining property of the Fatou set $\widehat{\mathbb{C}} \smallsetminus J$ for a rational function as follows. If V is a connected open subset of $\widehat{\mathbb{C}} \smallsetminus J$, show that the set of all limits of successive iterates $f^{\circ n}|_V$ as $n \to \infty$ is either (1) a finite set of constant maps from V into an attracting or parabolic periodic orbit, or (2) a compact one-parameter family of maps, consisting of all compositions

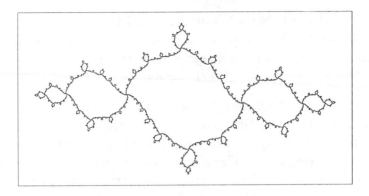

Figure 30. Julia set for $z \mapsto z^2 - 1 + 0.1i$. *(Problem 16-b.)*

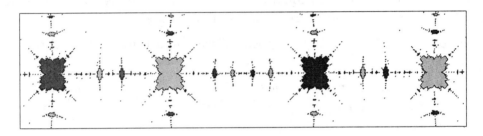

Figure 31. Julia set for $z \mapsto z + \sin(2\pi z)$. *(Problem 16-c.) Here, unlike all other Julia set pictures in these notes, the Julia set has been colored white.*

Figure 32. Julia set for $z \mapsto z + e^z - 1$. *(Problem 16-d.)*

$R_\theta \circ f^{\circ k}|_V$, with $k_0 \leq k < k_0 + p$. Here $f^{\circ k_0}$ is to be some fixed iterate with values in a rotation domain belonging to a cycle of rotation domains of period p , and R_θ is the rotation of this domain through angle θ .

Problem 16-b. Counting Components. If a quadratic polynomial map has an attracting fixed point or a parabolic fixed point of multiplier $\lambda = 1$, show that there is only one bounded Fatou component. (Figures 1a, 6, 20.) If it has an attracting cycle of period 2, show that there are three bounded components which map according to the pattern $U_1 \leftrightarrow U_0 \leftarrow U_1'$, and that the remaining bounded components are iterated preimages of U_1' where each set $f^{-n}(U_1')$ is made up of 2^n distinct components. Identify nine of these components in Figure 30. What is the corresponding description for a cycle of attracting or parabolic basins with period p (Figures 1d, 18), or for the case of a Siegel fixed point (Figures 22, 24)

Problem 16-c. Wandering Domains. Show that the transcendental map $f(z) = z + \sin(2\pi z)$ has one family of wandering domains $\{U_n\}$ with $f(U_n) = U_n + 1$ and one family $\{V_n\}$ with $f(V_n) = V_n - 1$. (Figure 31.)

Problem 16-d. A Baker Domain. Show that the map $f(z) = z + e^z - 1$ has a fully invariant Baker domain $U = f^{-1}(U)$. (Figure 32.)

Problem 16-e. Iterates of f^{-1} . Let $U \subset \widehat{\mathbb{C}}$ be a connected open set, and suppose there exists a smooth branch $g_k : U \to \widehat{\mathbb{C}}$ of $f^{-k}|_U$ for each $k \geq 1$. Show that the g_k form a normal family. If U contains a point of the Julia set, show that the norm $\|Dg_k\|$ of the first derivative, using the spherical metric, tends locally uniformly to zero. (Otherwise some subsequence of $\{g_k\}$ would converge to a non-constant limit $\hat{g}$, and the image $\hat{g}(U)$ would contain repelling periodic points $\cdots$.)

(Compare Lyubich 1986, p. 76. This approach, which goes back to Fatou, could be used to prove 16.5 without using 16.4.)

USING THE FATOU SET TO STUDY THE JULIA SET

§17. Prime Ends and Local Connectivity

Carathéodory's theory of "prime ends" is the basic tool for relating an open set of complex numbers to its complementary closed set. Let U be a simply connected subset of $\widehat{\mathbb{C}}$ such that the complement $\widehat{\mathbb{C}} \smallsetminus U$ is infinite. The Riemann Mapping Theorem asserts that there is a conformal isomorphism, which we write as

$$\psi : \mathbb{D} \xrightarrow{\cong} U .$$

In some cases, ψ will extend to a homeomorphism from the closed disk $\overline{\mathbb{D}}$ onto the closure $\overline{U}$. (Compare Figures 1a and 33a, together with 17.16) However, this is certainly not true in general, since the boundary ∂U may be an extremely complicated object. As an example, Figure 33b shows a region U such that one point of ∂U (with countably many short spikes sticking out) corresponds to a Cantor set of distinct points of the circle $\partial \mathbb{D}$. Figures 33c, 33d show examples for which an entire interval of points of ∂U corresponds to a single point of the circle. An effective analysis of the relationship between the compact set ∂U and the boundary circle $\partial \mathbb{D}$ was carried out by Carathéodory in 1913, and will be described here.

Finding Short Arcs. The main construction will be purely topological, but we first use analytic methods to prove several lemmas about the existence of short arcs. Let $I = (0, \delta)$ be an open interval of real numbers, and let $I^2 \subset \mathbb{C}$ be the open square, consisting of all $z = x + iy$ with $x, y \in I$. Suppose that we are given some conformal metric on I^2 of the form $\rho(z)|dz|$ where $\rho : I^2 \to (0, \infty)$ is a continuous strictly positive real valued function. (We do not assume that $\rho(z)$ is bounded.) By definition, the *area* of I^2 in this metric is the integral

$$\mathcal{A} = \iint_{I^2} \rho(x + iy)^2 \, dx \, dy ,$$

and the *length* of each horizontal line segment $y = \text{constant}$ is the integral

$$L(y) = \int_I \rho(x + iy) \, dx .$$

We will need the following. (Compare Appendix B.)

Lemma 17.1. Length-Area Inequality. *If the area $\mathcal{A}$ is finite, then the length $L(y)$ is finite for almost every height $y \in I$, and the average of $L(y)^2$ satisfies*

$$\frac{1}{\delta} \int_I L(y)^2 \, dy \leq \mathcal{A} . \tag{17 : 1}$$

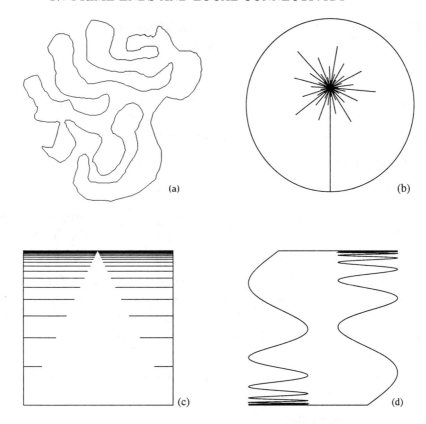

Figure 33. The boundaries of four simply connected regions in $\mathbb{C}$.

Proof. We will use the Schwarz Inequality in the form[*]

$$\left(\int_I f(x)g(x)\,dx\right)^2 \leq \left(\int_I f(x)^2\,dx\right)\left(\int_I g(x)^2\,dx\right), \qquad (17:2)$$

where f and g are square integrable real valued functions on I. Taking $f(x) = 1$ and $g(x) = \rho(x + iy)$, this yields

$$L(y)^2 \leq \delta \cdot \int_I \rho(x + iy)^2\,dx .$$

Integrating this inequality over y and dividing by δ, we obtain the required inequality $(17:1)$. If $\mathcal{A}$ is finite, it evidently follows that $L(y)$ is finite for y outside of a set of Lebesgue measure zero. □

For a "majority" of values of y, we can give a more precise upper bound as follows.

---

[*] If $\int f^2 = t^2 \int g^2$ where $t > 0$, then $(17:2)$ can be proved by manipulating the inequalities $\int (f \pm tg)^2 \geq 0$. The case where $\int f^2 = 0$ can be proved in a similar manner by letting t tend to zero.

Corollary 17.2. *The set S consisting of all $y \in I$ with $L(y) \leq \sqrt{2\mathcal{A}}$ has Lebesgue measure $\ell(S) > \ell(I)/2$.*

Proof. Evidently

$$\delta\,\mathcal{A} \geq \int_I L(y)^2 dy > \int_{I \smallsetminus S} \left(\sqrt{2\mathcal{A}}\right)^2 dy + \int_S 0 = 2\mathcal{A}\,\ell(I \smallsetminus S)\,,$$

and the conclusion follows since $\ell(I) = \delta$. $\square$

In the application, consider some univalent embedding

$$\eta : I^2 \overset{\cong}{\Longrightarrow} U \subset \hat{\mathbb{C}}\,.$$

Pulling the spherical metric from U back to I^2, we obtain a conformal metric of the form $\rho(z)|dz|$ on I^2. (Compare (2:4).) Evidently the area $\mathcal{A}$ of I^2 in this metric is at most equal to the area 4π of $\hat{\mathbb{C}}$.

Corollary 17.3. *Given such a univalent embedding of I^2 onto $U \subset \hat{\mathbb{C}}$, almost every horizontal line segment $y = $ constant in I^2 maps to a curve of finite spherical length; and more than half (in the sense of Lebesgue measure) of these horizontal line segments have spherical length at most $\sqrt{2\mathcal{A}}$, where $\mathcal{A}$ is the spherical area of U. Similar statements hold for vertical line segments $x = $ constant .*

The proof is immediate. $\square$

Now consider a simply connected open set $U \subset \hat{\mathbb{C}}$ with infinite complement, and some choice of conformal isomorphism $\psi : \mathbb{D} \to U$.

Theorem 17.4 (Fatou; Riesz and Riesz). *For almost every point $e^{i\theta}$ of the circle $\partial\mathbb{D}$ the radial line $r \mapsto re^{i\theta}$ maps under ψ to a curve of finite spherical length in U. In particular, the radial limit*

$$\lim_{r \nearrow 1} \psi(re^{i\theta}) \in \partial U$$

exists for Lebesgue almost every θ. However, if we fix any particular point $u_0 \in \partial U$, then the set of θ such that this radial limit is equal to u_0 has Lebesgue measure zero,

We will say briefly that almost every image curve $r \mapsto \psi(re^{i\theta})$ in U *lands* at some single point of ∂U, and that different values of θ almost always correspond to distinct landing points.

Remark. Fatou, in his thesis, showed that any bounded holomorphic function on $\mathbb{D}$ has radial limits in almost all directions, whether or not it is univalent. (See for example Hoffman, p. 38.) However the univalent case is all that we will need, and is easier to prove than the general theorem.

Proof of 17.4. The first half of 17.4 follows easily from 17.3, as follows. Let $\mathbb{H}^-$ be the left half-plane, consisting of all points $x + iy \in \mathbb{C}$ with $x < 0$. Map $\mathbb{H}^-$ onto $\mathbb{D} \smallsetminus \{0\}$ by the exponential map $x + iy \mapsto e^x e^{iy}$. Then the square

$$-2\pi < x < 0 , \qquad 0 \le y < 2\pi$$

in $\mathbb{H}^-$ maps under $\psi \circ \exp$ onto a neighborhood of the boundary in U. Almost every line $y = $ constant in $\mathbb{H}^-$ maps onto a curve of finite spherical length, which therefore tends to a well defined limit as $x \nearrow 1$.

If $U = \psi(\mathbb{D})$ is a bounded subset of $\mathbb{C}$, then a Theorem of F. and M. Riesz, as stated in §11 and proved in A.3 of Appendix A, asserts that any given radial limit can occur only for a set of directions $e^{i\theta}$ of measure zero. For any univalent ψ, we can reduce to the bounded case in two steps, as follows. First suppose that the image $\psi(\mathbb{D}) = U$ omits an entire neighborhood of some point z_0 of $\widehat{\mathbb{C}}$. Then by composing ψ with a fractional linear transformation which carries z_0 to ∞, we reduce to the bounded case. In general, $\psi(\mathbb{D})$ must omit at least two values, which we may take to be 0 and ∞. Then $\sqrt{\psi}$ can be defined as a single valued function which omits an entire open set of points; and we are reduced to the previous case. $\square$

Here is a topological complement. (Compare Figure 37, taking U to be the outside $\widehat{\mathbb{C}} \smallsetminus K$.)

Lemma 17.5. *If two different curves* $r \mapsto \psi(re^{i\theta_1})$ *and* $r \mapsto \psi(re^{i\theta_2})$ *land at the same point* $u_0 \in \partial U$, *then this point* u_0 *disconnects the boundary of* U.

Proof. These two curves, together with their landing point, form a Jordan curve Γ, which separates the sphere $\widehat{\mathbb{C}}$. Similarly, the angles θ_1 and θ_2 separate the circle $\mathbb{R}/2\pi\mathbb{Z}$ into two corresponding connected intervals. In particular, almost all angles in one of these two intervals correspond to curves which land in one connected component of $\widehat{\mathbb{C}} \smallsetminus \Gamma$, and almost all angles in the other interval correspond to curves which land in the other connected component. Since the landing points belong to the connected set ∂U, and since the separating curve Γ intersects ∂U only at u_0, this proves that the point u_0 separates ∂U. $\square$

Prime Ends. Next we descibe some constructions which depend only on the topology of the pair $(\overline{U}, \partial U)$ and not on conformal structure. We continue to assume that U is a simply connected open subset of the sphere $\widehat{\mathbb{C}}$, and that ∂U has more than one element.

Definition. By a *crosscut* (or "transverse arc") for the pair $(\overline{U}, \partial U)$ will be meant a subset $A \subset U$ which is homeomorphic to the open interval $(0, 1)$, such that the closure $\overline{A}$ is homeomorphic to a closed interval with only the two endpoints in ∂U.

Note that it is very easy to construct examples of crosscuts. For example we can start with any short line segment inside U and extend in both directions until it first hits the boundary.

Lemma 17.6. *Any crosscut A divides U into two connected components.*

Proof. The quotient space $\overline{U}/\partial U$, in which the boundary is identified to a point, is evidently homeomorphic to the 2-sphere. Since $\overline{A}$ corresponds to a Jordan curve in this quotient 2-sphere, the conclusion follows from the Jordan Curve Theorem. (See for example Munkres.) □

Either of the two connected components of $U \smallsetminus A$ will be called briefly a *crosscut neighborhood* $N \subset U$. Note that we can recover the crosscut A from such a crosscut neighborhood N since $\overline{A} = \overline{N} \cap \overline{U \smallsetminus N}$ and $A = \overline{A} \cap U$.

Main Definition. By a *fundamental chain* $\mathcal{N} = \{N_j\}$ in U will be meant a nested sequence

$$N_1 \supset N_2 \supset N_3 \supset \cdots$$

of crosscut neighborhoods $N_j \subset U$ such that the closures $\overline{A}_j$ of the corresponding crosscuts $A_j = U \cap \partial N_j$ are disjoint, and such that the diameter of $\overline{A}_j$ tends to zero as $j \to \infty$ (using the spherical metric). Two fundamental chains $\{N_j\}$ and $\{N_k'\}$ are *equivalent* if every N_j contains some N_k', and conversely every N_k' contains some N_j. An equivalence class $\mathcal{E}$ of fundamental chains is called a *prime end* for the pair $(\overline{U}, \partial U)$.

There are a number of possible minor variations on these basic definitions. (Compare Ahlfors 1973, D. Epstein, Ohtsuka.) The present version is fairly close to Carathéodory's original construction.

Note that only the crosscuts $\overline{A}_j$ are required to become small as $i \to \infty$. In examples such as Figure 33c and 33d, the crosscut neighborhood N_j may well have diameter bounded away from zero.

Definition. The intersection of the closures $\overline{N}_j \subset \overline{U}$ is called the *impression* of the fundamental chain $\{N_j\}$ or of the corresponding end $\mathcal{E}$.

Lemma 17.7. *For any fundamental chain $\{N_j\}$, the intersection of the open sets N_j is vacuous. However the impression $\cap \overline{N}_j$ is a non-vacuous compact connected subset of ∂U.*

Proof. For any $z \in U$ we will find a j with $z \notin N_j$. Choose a point $z_0 \in U \smallsetminus N_1$ and a path $P \subset U$ joining z_0 to z. If δ is the distance from the compact set P to ∂U, and if j is large enough so that the diameter of $\overline{A}_j$ is less than δ, then evidently $\overline{A}_j \cap P = \emptyset$, so A_j cannot disconnect z_0 from z. Since $z_0 \notin N_j$, it follows that $z \notin N_j$. Hence $\cap \overline{N}_j$ is a subset of ∂U. This set is clearly compact and non-vacuous. For the proof that it is connected, compare Problem 5-b. $\quad\square$

This impression may consist of a single point $z_0 \in \partial U$, as in Figures 33a and 33b. In this case we say that $\{N_j\}$ or $\mathcal{E}$ *converges* to the point z_0. Evidently the impression consists of a single point if and only if the diameter of N_j tends to zero as $j \to \infty$. However, in examples such as Figure 33c and 33d (as well as Problem 5-a and Figures 34, 36) the impression may well be a non-trivial continuum. Note also that two different prime ends may converge to the same point (Figures 33b, 33c), or more generally have the same impression.

We will say that two fundamental chains $\{N_j\}$ and $\{N'_k\}$ are *eventually disjoint* if $N_j \cap N'_k = \emptyset$ whenever both j and k are sufficiently large.

Lemma 17.8. *Any two fundamental chains $\{N_j\}$ and $\{N'_k\}$ in U are either equivalent or eventually disjoint.*

Proof. If $N_j \cap N'_k \neq \emptyset$ for all j and k, then we will show that for each j there is a k so that $N_j \supset N'_k$. We first show that every crosscut A'_k with k sufficiently large must intersect the neighborhood N_{j+1}. In fact we have assumed that every N'_k intersects N_{j+1}. Since $\cap N'_k = \emptyset$ by 17.7, it follows that the complement $U \smallsetminus N'_k$ must also intersect N_{j+1} for large k. Since N_{j+1} is connected, this implies that the common boundary A'_k must intersect N_{j+1}.

If no N'_k were contained in N_j, then every N'_k would intersect the complement $U \smallsetminus N_j$. An argument just like that above would then show that A'_k must intersect $U \smallsetminus N_j$ whenever k is large. But if A'_k intersects both $U \smallsetminus N_j$ and N_{j+1}, then it must cross both A_j and A_{j+1}. Hence its diameter must be greater than or equal to the distance between $\overline{A}_j$ and $\overline{A}_{j+1}$. This completes the proof, since it contradicts the hypothesis that the diameter of A'_k tends to zero. $\quad\square$

Now we will combine the topological and analytic arguments. We return to the study of a conformal isomorphism $\psi : \mathbb{D} \overset{\cong}{\longrightarrow} U \subset \widehat{\mathbb{C}}$.

Main Lemma 17.9. *Given any point $e^{i\theta}$ on the circle $\partial \mathbb{D}$, there exists a fundamental chain $\{N_j\}$ in $\mathbb{D}$ which converges to $e^{i\theta}$, and which maps under ψ to a fundamental chain $\{\psi(N_j)\}$*

in U .

Proof. We must construct the $N_1 \supset N_2 \supset \cdots$ in $\mathbb{D}$, converging to $e^{i\theta}$, so that the associated crosscuts A_j map to crosscuts in U which have disjoint closures, and which have diameters tending to zero. As in the proof of 17.4, we will make use of the exponential map $\exp : \mathbb{H}^- \to \mathbb{D} \smallsetminus \{0\}$, where $\mathbb{H}^-$ is the left half-plane. In fact we will actually construct crosscut neighborhoods N'_j in $\mathbb{H}^-$, converging to the boundary point $i\theta$, and then map to $\mathbb{D}$ by the exponential map. Each N'_j will be an open rectangle

$$-\epsilon < x < 0, \quad c_1 < y < c_2$$

in $\mathbb{H}^-$. Thus the corresponding crosscut $A'_j \subset \mathbb{H}^-$ will be made up of three of the four edges of this rectangle, and will have endpoints $0 + ic_1$ and $0 + ic_2$ in $\partial\mathbb{H}^-$. The construction will be inductive. Given $N'_1, \ldots, N'_{j-1}$ we first choose $\delta < 1/j$ which is small enough so that the square S_δ defined by the inequalities

$$-2\delta \le x < 0, \quad \theta - \delta \le y \le \theta + \delta$$

is contained in N'_{j-1}. Mapping S_δ into U by $\psi \circ \exp$, let $\mathcal{A}_\delta$ be the spherical area of its image. Evidently this area tends to zero as $\delta \to 0$. Using 17.2, we can choose constants c_1 and c_2 so that

$$\theta - \delta < c_1 < \theta < c_2 < \theta + \delta$$

and so that the horizontal line segments $y = c_k$ in S_δ map to curves of length at most $\sqrt{2\mathcal{A}_\delta}$ in $U \subset \widehat{\mathbb{C}}$. This will guarantee that the images of these line segments in U land at well defined points of ∂U as $x \nearrow 0$. We must also take care to see that these landing points are distinct from each other, and distinct from the endpoints of the crosscuts $\psi \circ \exp(A_h)$ with $h < j$. However, this does not pose any additional difficulty, in view of 17.4.

Finally, we must choose a vertical line segment $x = -\epsilon$ inside S_δ which also maps to a curve of length $\le \sqrt{2\mathcal{A}_\delta}$ in U. Setting

$$N'_j = (-\epsilon, 0) \times (c_1, c_2) \subset \mathbb{H}^-,$$

the inductive construction is complete. Mapping into $\mathbb{D}$, we obtain the required crosscut neighborhoods $N_j = \exp(N'_j) \subset \mathbb{D}$. $\square$

The inverse isomorphism $\psi^{-1} : U \to \mathbb{D}$ is much better behaved.

Corollary 17.10. *Any path $p : [0, 1) \to U$ which lands at a well defined point of ∂U maps under ψ^{-1} to a path in $\mathbb{D}$ which lands at a well defined point of $\partial\mathbb{D}$. Furthermore, paths which land at distinct points of ∂U map to paths which land at* *distinct points of* $\mathbb{D}$.

Proof. Let $e^{i\theta} \in \partial\mathbb{D}$ be any accumulation point of the path $t \mapsto \psi^{-1} \circ p(t)$ as $t \nearrow 1$. Choose some fundamental chain $\{N_j\}$ converging to $e^{i\theta}$ as in 17.9, so that the image under ψ is a fundamental chain $\{\psi(N_j)\}$ in U. We want to prove that $\psi^{-1} \circ p(t) \in N_j$ for all t which are sufficiently close to 1. Otherwise, for some j_0 we could find a sequence of points t_j converging to 1 so that $\psi^{-1} \circ p(t_j) \notin N_{j_0}$. Since $e^{i\theta}$ is an accumulation point of the path $\psi^{-1} \circ p$ in $\mathbb{D}$, this would imply that this path must pass through both of the crosscuts A_{j_0} and A_{j_0+1} infinitely often as $t \nearrow 1$. Hence the image path $p : [0, 1) \to U$ must pass through both $\psi(A_{j_0})$ and $\psi(A_{j_0+1})$ infinitely often. Since there is some positive distance between these crosscuts in U, this contradicts the hypothesis that $p(t)$ converges as $t \nearrow 1$.

If paths $p : [0, 1) \to U$ and $q : [0, 1) \to U$ landing at two distinct points of ∂U pulled back to paths $\psi^{-1} \circ p$ and $\psi^{-1} \circ q$ landing at a single point of $\partial\mathbb{D}$, then, choosing $\{N_j\}$ as above, each crosscut A_j with j large would cut both $\psi^{-1} \circ p$ and $\psi^{-1} \circ q$. Hence the image crosscut $\psi(A_j) \subset U$ would cut both p and q. As $j \to \infty$, the diameter of $\psi(A_j)$ tends to zero, while its intersection points with p and q tend to distinct points of ∂U. Evidently this is impossible. $\square$

Corollary 17.11. *Every fundamental chain* $\{N_k'\}$ *in* U *maps under* $\psi^{-1} : U \to \mathbb{D}$ *to a fundamental chain* $\{\psi^{-1}(N_k')\}$ *in* $\mathbb{D}$.

Proof. It follows from 17.10 that each $\psi^{-1}(N_k')$ is a crosscut neighborhood in $\mathbb{D}$, and that the closures of the associated crosscuts $\psi^{-1}(A_k')$ are disjoint. We must prove that the diameter of $\psi^{-1}(A_k')$ tends to zero as $j \to \infty$. Choose some accumulation point $e^{i\theta} \in \partial\mathbb{D}$ for these sets $\psi^{-1}(A_k')$, and choose a fundamntal chain $\{N_j\}$ in $\mathbb{D}$ which converges to $e^{i\theta}$ and which maps to a fundamental chain $\{\psi(N_j)\}$ in U. Then $N_j \cap \psi^{-1}(N_k') \neq \emptyset$ hence $\psi(N_j) \cap N_k' \neq \emptyset$ for all j and k. Therefore, by 17.8, the two fundamental chains $\{\psi(N_j)\}$ and $\{N_k'\}$ in U are equivalent. It follows that every N_j contains some $\psi^{-1}(N_k')$. Since the diameter of the entire neighborhood N_j clearly tends to zero as $j \to \infty$, the conclusion follows. $\square$

It follows easily from 17.9 and 17.11 that ψ induces a one-to-one corespondence between prime ends of $\mathbb{D}$ and prime ends of U. Furthermore the impression of any prime end of $\mathbb{D}$ is a single point of $\partial\mathbb{D}$, and each point of $\partial\mathbb{D}$ is the impression of one and only one prime end. We can express these facts in clearer form as follows.

Define the *Carathéodory compactification* $\hat{U}$ of U to be the disjoint union of U and the set consisting of all prime ends of U, with the following topology. For any crosscut neighborhood $N \subset U$ let $\tilde{N} \subset \hat{U}$ be the union of the set N itself, and the collection of all prime ends $\mathcal{E}$ which are represented by fundamental chains $\{N_j\}$ with $N_j \subset N$. These neighborhoods $\widetilde{N}$, together with the open subsets of U, form a basis for the required topology.

> **Theorem 17.12.** *The Carathéodory compactification* $\hat{\mathbb{D}}$ *of the open disk is canonically homeomorphic to the closed disk* $\overline{\mathbb{D}}$. *Furthermore, any conformal isomorphism* $\psi : \mathbb{D} \to U \subset \hat{\mathbb{C}}$ *extends uniquely to a homeomorphism from* $\overline{\mathbb{D}}$ *or* $\hat{\mathbb{D}}$ *onto* $\hat{U}$.

The proof is straightforward, and will be left to the reader. $\square$

Local connectivity. A Hausdorff space X is said to be *locally connected* if the following condition is satisfied:

(i) *Every point* $x \in X$ *has arbitrarily small connected (but not necessarily open) neighborhoods.*

Other equivalent conditions can be described as follows.

> **Lemma 17.13.** X *is locally connected if and only if:*
>
> (ii) *every* $x \in X$ *has arbitrarily small connected **open** neighborhoods, or*
>
> (iii) *every open subset of* X *is a union of connected open subsets.*
>
> *If* X *is compact metric, then an equivalent condition is that:*
>
> (iv) *for every* $\epsilon > 0$ *there exists* $\delta > 0$ *so that any two points of distance* $< \delta$ *are contained in a connected subset of* X *of diameter* $< \epsilon$.

Proof. It is easy to see that (iv) $\Rightarrow$ (i) $\Rightarrow$ (iii) $\Rightarrow$ (ii) $\Rightarrow$ (i). To show that (ii) $\Rightarrow$ (iv), let $\{Y_\alpha\}$ be the collection of all connected open sets of diameter $< \epsilon$, and let δ be the minimum of $\text{dist}(x, y)$ as (x, y) varies over the compact set $(X \times X) \smallsetminus \bigcup(Y_\alpha \times Y_\alpha)$, where $\delta > 0$ by (ii). $\square$

Remark. Sometimes it is important to study the situation around a single point $\mathbf{x} \in X$. There is no universally accepted usage, but it seems reasonable to say that X is *locally connected at* x if (i) is satisfied at the single point x, and *openly locally connected at* x if (ii) is satisfied. For the difference between these two requirements, see Problem 17-b.

Theorem 17.14 (Carathéodory). *A conformal isomorphism* $\psi : \mathbb{D} \xrightarrow{\cong} U \subset \hat{\mathbb{C}}$ *extends to a continuous map from the closed disk* $\overline{\mathbb{D}}$ *onto* $\overline{U}$ *if and only if the boundary* ∂U *is locally connected, or if and only if the complement* $\hat{\mathbb{C}} \smallsetminus U$ *is locally connected.*

Proof. If either ∂U or $\hat{\mathbb{C}} \smallsetminus U$ is locally connected, then we will show that for any fundamental sequence $\{N_j\}$ in U the impression $\cap \overline{N}_j$ consists of a single point. It will then follow easily that ψ extends continuously over the boundary of $\mathbb{D}$.

With ϵ and δ as in 17.13(iv), choose j large enough so that the crosscut $A_j = U \cap \partial N_j$ has diameter less than δ. It follows that the two endpoints of $\overline{A}_j$ have distance less than δ, and hence are contained in a compact connected set $Y \subset \hat{\mathbb{C}} \smallsetminus U$ of diameter less than ϵ. *Then the compact set* $Y \cup \overline{A}_j \subset \hat{\mathbb{C}}$ *separates* N_j *from* $U \smallsetminus \overline{N}_j$. For otherwise we could choose some smooth embedded arc $A' \subset \hat{\mathbb{C}}$ which is disjoint from $Y \cup \overline{A}_j$ and joins some point $x \in N_j$ to a point $y \in U \smallsetminus \overline{N}_j$. Taking A' together with a suitably chosen arc $A'' \subset U$ from x to y which cuts once across the crosscut A_j, we could construct a Jordan curve $A' \cup A''$ which separates the two endpoints of $\overline{A}_j$. Hence it would separate Y, which is impossible since Y was assumed connected.

This compact set $Y \cup \overline{A}_j$ has diameter less than $\epsilon + \delta$. If $\epsilon + \delta < \pi/2$, then one of the connected components of the complement $\hat{\mathbb{C}} \smallsetminus (Y \cup \overline{A}_j)$ contains an entire hemisphere, while all of the other connected components must have diameter less than $\epsilon + \delta$. If $\epsilon + \delta$ is also smaller than the diameter of $U \smallsetminus \overline{N}_1$, then it follows that $U \smallsetminus \overline{N}_j$ must be contained in the large component of $\hat{\mathbb{C}} \smallsetminus (Y \cup \overline{A}_j)$, hence N_j must have diameter less than $\epsilon + \delta$. Since ϵ and δ can be arbitrarily small, this proves that the impression $\cap \overline{N}_j$ can only be a single point. Using 17.9, it follows easily that ψ extends continuously over $\partial \mathbb{D}$.

To prove the converse statement, we need the following. Let us adopt the convention that all topological spaces are to be Hausdorff.

Lemma 17.15. *If* f *is a continuous map from a compact locally connected space* X *onto a Hausdorff space* Y, *then* Y *is also compact and locally connected.*

Proof. This image $f(X) = Y$ is certainly compact. Given any point $y \in Y$ and open neighborhood $N \subset Y$, we can consider the compact set $f^{-1}(y) \subset X$ with open neighborhood $f^{-1}(N)$. Let V_α range over all connected open subsets of $f^{-1}(N)$ which intersect $f^{-1}(y)$. Then the union $\cup f(V_\alpha)$ is a connected subset of N. It is also a neighborhood of

y, since it contains the open neighborhood $Y \smallsetminus f(X \smallsetminus \bigcup V_\alpha)$ of y. $\square$

The proof of 17.14 continues as follows. If ψ extends continuously to $\overline{\psi} : \overline{\mathbb{D}} \to \overline{U}$, then $\overline{\psi}$ maps the circle $\partial\mathbb{D}$ onto ∂U, so ∂U is locally connected. We must show that $\widehat{\mathbb{C}} \smallsetminus U$ is also locally connected. Here it is only necessary to consider the situation about a point $z_0 \in \partial U$, since $\widehat{\mathbb{C}} \smallsetminus U$ is clearly locally connected away from ∂U. Choose an arbitrarily small connected neighborhood N of z_0 within ∂U, and then choose ϵ so that the ball of radius ϵ about z intersected with ∂U is contained in N. The union of N and the ball of radius ϵ about z within $\widehat{\mathbb{C}} \smallsetminus U$ is then the required small connected neighborhood. $\square$

Combining this theorem with Lemma 17.5, we obtain the following.

Theorem 17.16. *If the boundary of U is a Jordan curve, then $\psi : \mathbb{D} \cong U$ extends to a homeomorphism from the closed disk $\overline{\mathbb{D}}$ onto the closure $\overline{U}$.*

Proof. If ∂U is a Jordan curve, that is a homeomorphic image of the circle, then we certainly have a continuous extension $\psi : \overline{\mathbb{D}} \to \overline{U}$ by 17.14. Since a Jordan curve cannot be separated by any single point, it follows from 17.5 that this extension is one-to-one, and hence is a homeomorphism. $\square$

Definitions. The space X is *path connected* if there exists a continuous map from the unit interval $[0, 1]$ into X which joins any two given points, and *arcwise connected* if there is a topological embedding of $[0, 1]$ into X which joins any two given distinct points. It is *locally path connected* if every point has arbitrarily small path connected neighborhoods.

To conclude this section, we prove two well known results. In general, a connected space need not be path connected (Figure 33d). However:

Lemma 17.17. *If a compact metric space X is locally connected, then it is locally path connected.*

It follows easily that every connected component is path connected. Furthermore:

Lemma 17.18. *If a Hausdorff space is path connected, then it is necessarily arcwise connected.*

Proof of 17.17. Let X be compact metric and locally connected. Given $\epsilon > 0$, it follows from 17.13 that we can choose a sequence of numbers $\delta_n > 0$ so that any two points with distance $\mathrm{dist}(x, x') < \delta_n$ are contained in a connected set of diameter less than $\epsilon/2^n$. *We will prove that any two points $x(0)$ and $x(1)$ with distance $\mathrm{dist}(x(0), x(1)) < \delta_0$ can be joined by a path of diameter at most 4ϵ.*

The plan of attack is as follows. We will choose a sequence of denominators $1 = k_0 < k_1 < k_2 < \cdots$, each of which divides the next, by induction. Also, for each fraction of the form i/k_n between 0 and 1 we will choose an intermediate point $x(i/k_n)$ satisfying the following condition: *If $|i/k_n - j/k_{n+1}| \leq 1/k_n$ then the distance between $x(i/k_n)$ and $x(j/k_{n+1})$ must be less than $\epsilon/2^n$*. Furthermore, the distance between $x(i/k_n)$ and $x((i+1)/k_n)$ will be less than δ_n. The inductive construction follows. Given $x(i/k_n)$ and $x((i+1)/k_n)$, choose some connected set C which contains both and has diameter less than $\epsilon/2^n$. The points $x(i/k_n)$ and $x((i+1)/k_n)$ can be joined within C by a finite chain of points so that two consecutive points have distance less than δ_{n+1}. Taking k_{n+1} to be a suitably large multiple of k_n, we can evidently choose the required points $x(j/k_{n+1})$ for $i/k_n < j/k_{n+1} < (i+1)/k_n$ from this chain, allowing duplications if necessary. Thus we may assume that $x(r)$ has been defined inductively for a dense set of rational numbers $r = i/k_n$ in the unit interval.

Next we will prove that this densely defined correspondence $r \mapsto x(r)$ is uniformly continuous. Let r and r' be any two rational numbers for which $x(r)$ and $x(r')$ are defined. If $|r - r'| \leq 1/k_n$, then we can choose i/k_n so that both $|r - i/k_n|$ and $|r' - i/k_n|$ are at most $1/k_n$. It follows easily that

$$\text{dist}\Big(x(r), \, x(i/k_n)\Big) < \epsilon/2^n + \epsilon/2^{n+1} + \cdots,$$

and similarly for $x(r')$. Hence $\text{dist}(x(r), x(r')) < 4\epsilon/2^n$. This proves uniform continuity; and it follows that there is a unique continuous extension $t \mapsto x(t)$ which is defined for all $t \in [0, 1]$. In this way, we have constructed the required path of diameter at most 4ϵ from $x(0)$ to $x(1)$. Thus X is locally path connected, as required. □

Proof of 17.18. Let $f = f_0 : [0, 1] \to X$ be any continuous path with $f(0) \neq f(1)$. We must construct an embedded arc $A \subset X$ from $f(0)$ to $f(1)$. Choose a closed subinterval $I_1 = [a_1, b_1] \subset [0, 1]$ whose length $0 \leq \ell(I_1) = b_1 - a_1 < 1$ is as large as possible, subject to the condition that $f(a_1) = f(b_1)$. Now, among all subintervals of $[0, 1]$ which are disjoint from I_1, choose an interval $I_2 = [a_2, b_2]$ of maximal length subject to the condition $f(a_2) = f(b_2)$. Continue this process inductively, constructing disjoint subintervals of maximal lengths $\ell(I_1) \geq \ell(I_2) \geq \cdots \geq 0$ subject to the condition that f is constant on the boundary of each I_j.

Let $\alpha : [0, 1] \to X$ be the unique map which takes the constant value $\alpha(I_j) = f(\partial I_j)$ on each of these closed intervals I_j, and which coincides with f outside of these subintervals. Then it is easy to check that α is continuous, and that for each point $x \in \alpha([0, 1])$ the preimage $\alpha^{-1}(x) \subset$

$[0,1]$ is a (possibly degenerate) closed interval of real numbers. Note that the image $A = \alpha([0,1]) \subset X$ can be totally ordered by specifying that $\alpha(s) \ll \alpha(t)$ if and only if $\alpha(s)$ and $\alpha(t)$ are distinct points with $s < t$.

A homeomorphism h between the interval $[0,1]$ and the set A can be constructed as follows. Choose a countable dense subset $\{t_1, t_2, \ldots\}$ in the open interval $(0,1)$ and a countable dense subset $\{a_1, a_2, \ldots\}$ in A, excluding the endpoints $\alpha(0)$ and $\alpha(1)$. Now construct a one-to-one correspondence $i \mapsto j(i)$ by induction: Let $j(1) = 1$, and if $j(1), j(2), \ldots, j(i-1)$ have already been chosen let $j(i)$ be the smallest positive integer which is distinct from $j(1), \ldots, j(i-1)$, and which satisfies the condition that

$$t_h < t_i \quad \Longleftrightarrow \quad a_{j(h)} \ll a_{j(i)}$$

for $h < i$. The required homeomorphism $h : [0,1] \to A$ is now defined by mapping each Dedekind cut in $\{t_i\}$ to the corresponding Dedekind cut in $\{a_j\}$, so that $h(t_i) = a_{j(i)}$ and

$$t < t_i \quad \Longleftrightarrow \quad h(t) \ll h(t_i) . \qquad \square$$

Problem 17-a. With $\psi : \mathbb{D} \xrightarrow{\cong} U$ as in 17.9, let $p_t : [0,1) \to U$ be a one-parameter family of paths in U, all landing at the same point of ∂U. Show that the paths $\psi^{-1} \circ p_t$ all land at the same point of $\partial \mathbb{D}$.

Problem 17-b. Let $X \subset \mathbb{C}$ be the compact connected set which is obtained from the unit interval $[0,1]$ by drawing line segments from 1 to the points $\frac{1}{2}(1 + i/n)$ for $n = 1, 2, 3, \ldots$ and then adjoining the successive images of this configuration under the map $z \mapsto z/2$. (Figure 34.) Show that X is locally connected at the origin, but not openly locally connected.

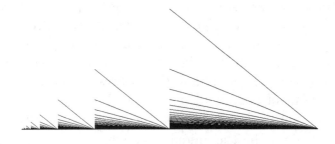

Figure 34. The witch's broom.

§18. Polynomial Dynamics: External Rays

First recall some definitions from §9. Let $f : \widehat{\mathbb{C}} \to \widehat{\mathbb{C}}$ be a monic polynomial map of degree $n \geq 2$, say

$$f(z) = z^n + a_{n-1}z^{n-1} + \cdots + a_1 z + a_0 \ .$$

Then f has a superattracting fixed point at infinity. In particular, it is not difficult to find a constant c_f so that every point z in the neighborhood $|z| > c_f$ of infinity belongs to the basin of attraction $\mathcal{A}(\infty)$. The complement of the basin $\mathcal{A}(\infty)$, that is the set of all points $z \in \mathbb{C}$ with bounded forward orbit under f, is called the *filled Julia set* $K = K(f)$. This filled Julia set is always a compact subset of the plane, consisting of the Julia set J together with the bounded components (if any) of the complement $\mathbb{C} \smallsetminus J$. These bounded components are all simply connected (compare Problem 15-a or the proof of 9.4), and the Julia set J is equal to the topological boundary ∂K. Throughout this section we will assume the following.

Standing Hypothesis. *The Julia set J is connected, or equivalently the filled Julia set K is connected.*

Then by 9.5 the complement $\mathbb{C} \smallsetminus K$ is conformally isomorphic to $\mathbb{C} \smallsetminus \overline{\mathbb{D}}$ under the Böttcher isomorphism

$$\phi : \mathbb{C} \smallsetminus K \xrightarrow{\cong} \mathbb{C} \smallsetminus \overline{\mathbb{D}}$$

which conjugates the map f outside K to the n-th power map $w \mapsto w^n$ outside the closed unit disk, with $\phi(z)$ asymptotic to the identity map at infinity. The function $G : \mathbb{C} \to \mathbb{R}$ defined by

$$G(z) = \begin{cases} \log|\phi(z)| > 0 & \text{for} \quad z \in \mathbb{C} \smallsetminus K \\ 0 & \text{for} \quad z \in K \end{cases}$$

is called the *Green's function* for K. (Compare 9.6.) Note the identity

$$G(f(z)) = nG(z) \ .$$

Each locus $G^{-1}(c) = \{z \ ; \ G(z) = c\}$ with $c > 0$ is called an *equipotential curve* around the filled Julia set K. Note that f maps each equipotential $G^{-1}(c)$ to the equipotential $G^{-1}(nc)$ by a n-to-one fold covering map. The orthogonal trajectories

$$\{z \ ; \ \arg(\phi(z)) = \text{constant}\}$$

of the family of equipotential curves are called *external rays* for K. We will use the notation $R_t \subset \mathbb{C} \smallsetminus K$ for the external ray with angle t, where now we measure angle as a fraction of a full turn, so that $t \in \mathbb{R}/\mathbb{Z}$. By

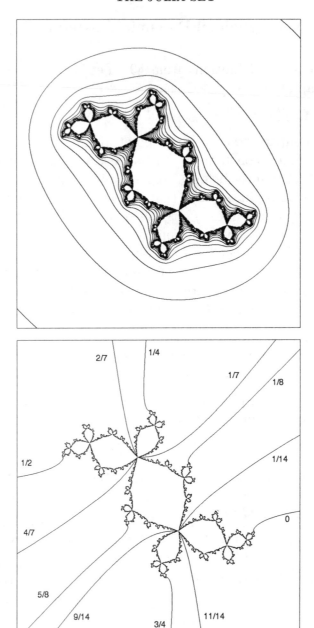

Figure 35. Julia set for the "Douady rabbit"

$$z \; \mapsto \; z^2 - .12256 + .74486i \; .$$

In the top figure, some equipotentials of the form $G = 2^n G_0$ have been drawn in. The lower figure shows several periodic and pre-periodic external rays.

definition, R_t is the image under the inverse Böttcher map ϕ^{-1} of the half-line consisting of all products $re^{2\pi it} \in \mathbb{C} \smallsetminus \overline{\mathbb{D}}$, with $r > 1$. Note the identity

$$f(R_t) = R_{nt}.$$

In particular, if the angle $t \in \mathbb{R}/\mathbb{Z}$ is periodic under multiplication by n, then the ray R_t is periodic. For example if $n^p t \equiv t \pmod{\mathbb{Z}}$, then it follows that $f^{\circ p}$ maps the ray R_t onto itself.

Now consider the limit

$$\gamma(t) = \lim_{r \searrow 1} \phi^{-1}(re^{2\pi it}).$$

Whenever this limit exists, we will say that the ray R_t *lands* at the point $\gamma(t)$, which necessarily belongs to the Julia set $J = \partial K$.

Lemma 18.1. *If the ray R_t lands at a single point $\gamma(t)$ of the Julia set, then the ray R_{nt} lands at the point $\gamma(nt) = f(\gamma(t))$. Furthermore each of the n rays of the form $R_{(t+j)/n}$ lands at one of the points in $f^{-1}(\gamma(t))$, and every point in $f^{-1}(\gamma(t))$ is the landing point of at least one such ray.*

Proof. If $z \in J$ is not a critical point, then f maps a neighborhood N of z diffeomorphically onto a neighborhood N' of $f(z)$, carrying any ray $R_s \cap N$ to $R_{ns} \cap N'$. Thus if R_s lands at z then R_{ns} lands at $f(z)$, while if R_t lands at $f(z)$ then for some uniquely determined s of the form $(t+j)/n$ the ray R_s must land at z. If z is a critical point, the situation is similar, except that N maps to N' by a branched covering, so that each ray landing at $f(z)$ is covered by two or more rays landing at z. $\square$

In particular, if the ray R_t is periodic of period $p \geq 1$, and if R_t lands at a point $\gamma(t)$, then it follows that $\gamma(t)$ is a periodic point of f with period dividing p.

Fatou showed that most rays do land, and the Riesz brothers showed that distinct angles usually correspond to distinct landing points. More precisely, applying 17.4 to the basin of infinity $\mathcal{A}(\infty) \subset \widehat{\mathbb{C}}$, we have the following corollary.

Theorem 18.2. Most rays land. *For all $t \in \mathbb{R}/\mathbb{Z}$ outside of a set of measure zero, the ray R_t has a well defined landing point $\gamma(t) \in J(f)$. For each fixed $z_0 \in J$, the set of t with $\gamma(t) = z_0$ has measure zero.*

However, it is definitely not true that rays land in all cases. Using Carathéodory's work, we can give a precise criterion.

Theorem 18.3. Landing criterion. *For any given f with connected Julia set, the following four conditions are equivalent.*

- *Every external ray R_t lands at a point $\gamma(t)$ which depends continuously on the angle t .*

- *The Julia set J is locally connected.*

- *The filled Julia set K is locally connected.*

- *The inverse Böttcher map $\phi^{-1} : \mathbb{C} \smallsetminus \overline{\mathbb{D}} \to \mathbb{C} \smallsetminus K$ extends continuously over the boundary $\partial \mathbb{D}$, mapping $e^{2\pi i t} \in \partial \mathbb{D}$ to $\gamma(t) \in J(f)$.*

Furthermore, whenever these conditions are satisfied, the resulting map $\gamma : \mathbb{R}/\mathbb{Z} \to J(f)$ satisfies the semiconjugacy identity

$$\gamma(nt) = f(\gamma(t)) ,$$

and maps the circle $\mathbb{R}/\mathbb{Z}$ onto the Julia set $J(f)$.

Definition. This map γ from $\mathbb{R}/\mathbb{Z}$ onto J will be called the *Carathéodory semiconjugacy* associated with the locally connected polynomial Julia set J .

Proof of 18.3. First suppose that $\gamma : \mathbb{R}/\mathbb{Z} \to J$ is defined and continuous. Then the image $\gamma(\mathbb{R}/\mathbb{Z})$ is certainly a non-vacuous compact subset of J . Starting with an arbitrary point, say $\gamma(0)$, in this image, we see inductively, using 18.1, that all iterated preimages also belong to $\gamma(\mathbb{R}/\mathbb{Z})$. Therefore by 4.10, the image $\gamma(\mathbb{R}/\mathbb{Z})$ is the entire Julia set J , and by 17.15, J is locally connected. The remaining statements in 18.3 now follow immediately from Carathéodory's Theorem 17.14, applied to the conformal isomorphism $\widehat{\mathbb{C}} \smallsetminus \overline{\mathbb{D}} \to \widehat{\mathbb{C}} \smallsetminus K$.   □

Remark. A priori, it is possible that every external ray may land even if K is not locally connected. An example of a compact set (but not a filled Julia set) with this property is the "symmetric comb", shown in Figure 36. It consists of the lines $[-1, 1] \times \{\pm c^n\}$ (here $c = .75$), together with the axes $[-1, 1] \times \{0\}$ and $\{0\} \times [-1, 1]$. Evidently, in such an example, the associated landing point function $t \mapsto \gamma(t)$ cannot be continuous.

Here is another result, which follows immediately from Carathéodory's Theorem 17.16.

Corollary 18.4. Simple closed curves. *The Julia set J is a simple closed curve if and only if γ maps $\mathbb{R}/\mathbb{Z}$ homeomorphically onto J .*

For examples, see Figures 1a, 6, 10, 20.

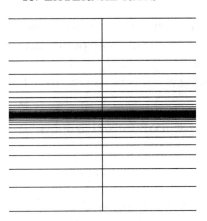

Figure 36. A symmetric comb.

Following is a basic result due to Sullivan and Douady. (Sullivan 1983. See also Lyubich 1986, p. 85.)

Theorem 18.5. Locally connected Julia sets. *If the Julia set J of a polynomial map f is locally connected, then every periodic point in J is either repelling or parabolic. Furthermore every cycle of Siegel disks for f contains at least one critical point on its boundary.*

Recall from §11 that a *Cremer point* can be characterized as a periodic point which belongs to the Julia set but is neither repelling nor parabolic. Thus the following is a completely equivalent statement.

Corollary 18.6. Non locally connected Julia sets. *If f is a polynomial map with a Cremer point, or with a cycle of Siegel disks whose boundary contains no critical point, then the Julia set $J(f)$ is not locally connected.*

Remark. It is essential for these results that f be a polynomial. For example a rational map with a Cremer point may well have locally connected Julia set. (Compare [Roesch].) In fact the Julia set can even be the entire Riemann sphere (using Shishikura, 1987). However the following supplementary statement holds even for non-polynomials.

Lemma 18.7. Siegel disk boundaries. *If a rational map f has a Siegel disk $\Delta = f(\Delta)$ such that either the boundary $\partial\Delta$ or the Julia set $J(f)$ is locally connected, then $\partial\Delta$ must be a simple closed curve, and f restricted to $\partial\Delta$ must be topologically conjugate to an irrational rotation. In particular, there can be no periodic points in the boundary.*

Examples of Cremer points were constructed in §11, and examples of Siegel disks with no boundary critical point have been given by Herman (1986; compare Douady 1987). However, I know of no example of a Siegel disk which has a boundary periodic point, or which is not bounded by a simple closed curve.

Proof of 18.7. Choose a conformal isomorphism $\psi : \mathbb{D} \to \Delta$ satisfying the conjugacy identity

$$\psi(\rho w) = f\big(\psi(w)\big),$$

where ρ has the form $e^{2\pi i \alpha}$ with $\alpha \in \mathbb{R} \setminus \mathbb{Q}$. Arguing as in 17.14, we see that ψ extends to a continuous map $\Psi : \overline{\mathbb{D}} \to \overline{\Delta}$ which must satisfy this same identity. But this implies that $\partial \mathbb{D}$ maps homeomorphically onto $\partial \Delta$. For otherwise, if $\Psi(w_0) = \Psi(u w_0)$ for some $u \neq 1$, $|u| = 1$, then it would follow that $\Psi(\rho^k w_0) = \Psi(\rho^k u w_0)$ for $k = 1, 2, \ldots$, and hence that $\Psi(w) = \Psi(u w)$ for all w on the unit circle. If the group of all such u were dense on the circle, then $\Psi(\partial \mathbb{D})$ would be a single point, which is impossible. On the other hand, if this group were generated by some root of unity, then the result of gluing a neighborhood of w_0 in the boundary of $\overline{\mathbb{D}}$ to a neighborhood of $u w_0$ would be a non-orientable surface embedded in $\widehat{\mathbb{C}}$, which is also impossible. This contradiction proves 18.7. $\square$

The proof of 18.5 will be based on the following. Let z_0 be a fixed point in the Julia set J. If J is locally connected, then $\gamma : \mathbb{R}/\mathbb{Z} \to J$ is continuous and onto, hence the set $X = \gamma^{-1}(z_0)$ consisting of all angles t such that R_t lands at z_0 is a non-vacuous compact subset of the circle. We claim that the n-tupling map $t \mapsto nt$ carries X homeomorphically onto itself. In fact the fixed point z_0 certainly cannot be a critical point of f, since it lies in the Julia set, so f maps a small neighborhood of z_0 diffeomorphically onto a small neighborhood of z_0, carrying external rays landing at z_0 bijectively to external rays landing at z_0.

Lemma 18.8. *Let $n \geq 2$ be an integer, and let $X \subset \mathbb{R}/\mathbb{Z}$ be a compact set which is carried homeomorphically onto itself by the map $t \mapsto nt \pmod{\mathbb{Z}}$. Then X is finite.*

Proof. In fact we will prove the following more general statement. Let X be a compact metric space with distance function $\mathrm{dist}(x, y)$, and let $h : X \to X$ be a homeomorphism which is *expanding* in the following sense: There should exist numbers $\epsilon > 0$ and $k > 1$ so that

$$\mathrm{dist}(h(x), h(y)) \geq k \, \mathrm{dist}(x, y)$$

whenever $\mathrm{dist}(x, y) < \epsilon$. Then we will show that X is finite. Evidently this hypothesis is satisfied in the situation of 18.8, so this argument will prove the Lemma.

Since $h^{-1} : X \to X$ is uniformly continuous, we can choose $\delta > 0$ so that $\mathrm{dist}(x, y) < \epsilon$ whenever $\mathrm{dist}(h(x), h(y)) < \delta$. But this implies that $\mathrm{dist}(x, y) < \delta/k$. Since X is compact, we can choose some finite number, say m, of balls of radius δ which cover X. Applying h^{-p}, we obtain m balls of radius δ/k^p which cover X. Since p can be arbitrarily large, this proves that X can have at most m distinct points. $\square$

The proof of 18.5 will also require the following. We have shown that the set of external rays landing on the fixed point z_0 is finite, and maps bijectively to itself under f. Hence the angles of these rays must be periodic under multiplication by the degree n. Replacing f by some iterate if necessary, we may assume that these angles are actually fixed, $nt \equiv t \pmod{\mathbb{Z}}$, so that $f(R_t) = R_t$.

Lemma 18.9. *If a fixed ray $R_t = f(R_t)$ lands at z_0, then z_0 is either a repelling or a parabolic fixed point.*

(Compare 18.10 below.) This lemma is an immediate corollary of the Snail Lemma of §16. First note that each equipotential

$$\{z \in \mathbb{C} \; ; \; G(z) = \text{constant} > 0\}$$

intersects the ray R_t in a single point. Hence we can parametrize R_t as the image of a topological embedding $p : \mathbb{R} \to \mathbb{C} \smallsetminus K$, which maps each $s \in \mathbb{R}$ to the unique point $z \in R_t$ with $\log G(z) = s$. (In fact s can be identified with the Poincaré arclength parameter along R_t.) Since $G(f(z)) = n\, G(z)$, we have $f(p(s)) = p(s + \log n)$, and it follows from 16.3 that the landing point

$$z_0 = \gamma(t) = \lim_{s \to -\infty} p(s)$$

is indeed a repelling or parabolic fixed point. $\square$

Proof of 18.5. If z_0 is a fixed point in a locally connected Julia set, then the preceding two lemmas and the accompanying discussion show that z_0 must be a repelling or parabolic point. the extension to periodic points is straightforward. Now consider a fixed Siegel disk $\Delta = f(\Delta)$. Since $\gamma : \mathbb{R}/\mathbb{Z} \to J$ is continuous, it follows that the set $X = \gamma^{-1}(\partial \Delta) \subset \mathbb{R}/\mathbb{Z}$ is compact, and this set is clearly infinite. Hence by 18.8, the map $t \mapsto nt$ on X cannot be bijective. On the other hand, we know by 18.7 that f maps $\partial \Delta$ homeomorphically onto itself, and it follows that the n-tupling map carries X onto itself. Therefore, there must be two distinct rays R_{t_1} and R_{t_2} landing on $\partial \Delta$ with $f(R_{t_1}) = f(R_{t_2})$. Since $f|_{\partial \Delta}$ is one-to-one, these two rays must land at a common point $\gamma(t_1) = \gamma(t_2)$, and evidently this common landing point must be a critical point of f. This completes the proof of 18.5 for a Siegel disk of period one, and the extension to higher periods is straightforward. □

Definition. An external ray R_t is called *rational* if its angle $t \in \mathbb{R}/\mathbb{Z}$ is rational; and *periodic* if t is periodic under multiplication by the degree n so that $n^p t \equiv t \pmod{1}$ for some $p \geq 1$.

Note that R_t is eventually periodic under multiplication by n if and only if t is rational, and is periodic if and only if the number t is rational with denominator relatively prime to n. (If t is rational with denominator d, then the successive images of R_t under f have angles nt, $n^2 t$, $n^3 t$, ... $(\mod \mathbb{Z})$ with denominators dividing d. Since there are only finitely many such fractions modulo $\mathbb{Z}$, this sequence must eventually repeat. In the special case where d is relatively prime to n, the fractions with denominator d are permuted under multiplication by n modulo $\mathbb{Z}$, so the landing point $\gamma(t)$ is actually periodic.)

We continue to assume that K is connected. However, we will not assume local connectivity.

Theorem 18.10. Rational Rays Land. *Every periodic external ray lands at a periodic point which is either repelling or parabolic. If t is rational but not periodic, then the ray R_t lands at a point which is eventually periodic but not periodic.*

(Compare Douady and Hubbard 1984-85.) The converse result, due to Douady, is more difficult. (Compare Petersen 1991, Hubbard 1993.)

Theorem 18.11.   Repelling and Parabolic Points are Landing Points. *Every repelling or parabolic periodic point is the landing point of at least one periodic ray.*

The following supplementary statement is much easier to prove.

Lemma 18.12. *If one periodic ray lands at the point z_0, then only finitely many rays land at z_0, and these rays are all periodic of the same period (which may be larger than the period of z_0).*

As an example, Figure 37 shows the Julia sets for the cubic maps $f(z) = z^3 - 3z/4 + \sqrt{-7}/4$ and $g(z) = z^3 - iz^2 + z$. In the left hand example, the 0 and 1/2 rays land at distinct fixed points, the 1/8, 1/4, 3/8 and 3/4 rays land at the third fixed point, while the 5/8 and 7/8 rays land on a period two orbit. In the right hand example, the 0 and 1/2 rays

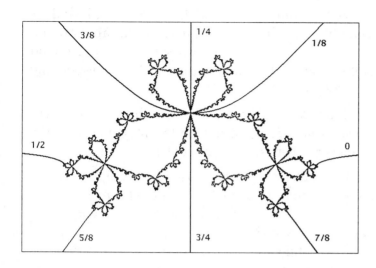
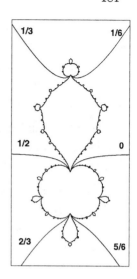

Figure 37. Julia sets for $z \mapsto z^3 - .75z + \sqrt{-7}/4$ and $z \mapsto z^3 - iz^2 + z$
with some external rays indicated.

must both land at the parabolic fixed point $z = 0$, since the remaining
fixed point at $z = i$ is superattracting and hence does not belong to the
Julia set. The $1/6, 1/3, 2/3$ and $5/6$ rays have denominator divisible by
3, and therefore land at preperiodic points: In fact, these four rays land at
the two disjoint pre-images of zero. (The analogous discussion for Figure
35 will be left to the reader.)

In the parabolic case, we can sharpen these statements as follows.

Theorem 18.13. The Parabolic Case. *If the multiplier at
a parabolic fixed point z_0 is a q-th root of unity $(q \geq 1)$, then
every ray which lands at z_0 has period q. For every repelling
petal P at z_0, there is at least one ray landing at z_0 through
the petal P.*

Example 1. Consider the cubic map $g(z) = z^3 - iz^2 + z$ of Figure
37. The parabolic fixed point $z = 0$ has multiplier $\lambda = 1$, so there is
only one repelling petal, yet two distinct rays R_0 and $R_{1/2}$ land at this
point. Figure 15 shows a similar example, with three repelling petals but
four fixed landing rays.

Example 2. Now consider the map $f(z) = z^2 + \exp(2\pi i \cdot 3/7)z$ of
Figure 18. Here the multiplier is a seventh root of unity, and there are
seven repelling petals about the origin. Hence there must be at least seven
external rays landing at the origin, and their angles must be fractions with
denominator $127 = 2^7 - 1$, so as to be periodic of period 7. In fact a

little experimentation shows that only the ray with angle $21/127$ and its successive iterates under doubling modulo 1 will fit in the right order around the origin. (Compare Goldberg.) Thus there are just seven rays which land at zero, one in each repelling petal. The numerators of the corresponding angles are 21, 42, 84, 41, 82, 37, 74.

The proofs begin as follows. We will continue to use the notation $\gamma(t)$ for landing points. However, since K need not be locally connected, the function $\gamma(t)$ need not be everywhere defined or continuous.

Proof of 18.12. First consider the special case of a fixed ray $R_{t_0} = f(R_{t_0})$. In other words, suppose that t_0 is a number of the form $j/(n-1)$, so that $t_0 \equiv nt_0 \pmod{\mathbb{Z}}$. If R_{t_0} lands at z_0, then clearly $f(z_0) = z_0$. Let X be the set of all angles x such that the ray R_x lands at z_0. Since f maps a neighborhood of z_0 diffeomorphically onto a neighborhood of z_0, preserving the cyclic order of the rays which land at z_0, it follows that the n-tupling map carries X injectively into itself preserving cyclic order.

For every $x \in X$, we must show that R_x is also mapped onto itself by f. This will imply that there are at most $n-1$ rays landing at z_0, and that they all have the same period $p = 1$. Otherwise, if $nx \not\equiv x \pmod{\mathbb{Z}}$, then it follows that $x \not\equiv t_0 \pmod{\mathbb{Z}}$, and hence that $kx \not\equiv t_0$ for all k. Define the sequence $x_0, x_1, \ldots$ of representative points for the orbit of x within the interval $(t_0, t_0 + 1)$ by the congruence

$$x_k \equiv n^k x \pmod{\mathbb{Z}} \qquad \text{with} \quad t_0 < x_k < t_0 + 1.$$

Suppose, to fix our ideas, that $t_0 < x_0 < x_1 < t_0 + 1$. Since the n-tupling map preserves cyclic order, it follows that the images of t_0, x_0, and x_1 must satisfy $t_0 < x_1 < x_2 < t_0 + 1$. Continuing inductively, it follows that $t_0 < x_0 < x_1 < x_2 < \cdots < t_0 + 1$. Thus the x_k must converge to some angle $\hat{x}$, which is necessarily a fixed point for the map $t \mapsto nt \pmod{\mathbb{Z}}$. But this is impossible, since this map has only strictly repelling fixed points.

Now suppose that the smallest period of a ray R_t landing at z_0 is $p > 1$. Replacing f by the iterate $g = f^{\circ p}$, the argument above shows that every ray which lands at z_0 is carried into itself by g, and hence has period $\leq p$ under f. This proves that the period is exactly p. $\square$

Proof of Theorem 18.10. We will make use of the Poincaré metric on $\mathbb{C} \smallsetminus K \cong \mathbb{C} \smallsetminus \overline{\mathbb{D}}$. Note that the map f is a local isometry for this metric. In fact the universal covering of $\mathbb{C} \smallsetminus \overline{\mathbb{D}}$ is isomorphic to the right half-plane $\{w = u + iv \; ; \; u > 0\}$ under the exponential map. Here the real part u of w corresponds to the Green's function G on $\mathbb{C} \smallsetminus K$. The map f on $\mathbb{C} \smallsetminus K$ corresponds to the n-th power map on $\mathbb{C} \smallsetminus \overline{\mathbb{D}}$, which corresponds to the Poincaré isometry $w \mapsto nw$ of the right half-plane. Note also that each external ray corresponds to a horizontal half-line $v = $ constant . The Poincaré arclength $\int |dw|/u$ reduces to $\int du/u = \int d \log u$ along each such half-line.

Again, first consider the case of a fixed ray $f(R_t) = R_t$. As in the proof of 18.9, we can introduce the parameter $s = \log G(z)$ along this ray, so that R_t is the image of a path $p : \mathbb{R} \to \mathbb{C} \smallsetminus K$, where

$$f(p(s)) \; = \; p(s + \log n) \ .$$

Thus $R_t = p(\mathbb{R})$ is the union of path segments

$$I_k = p\big([k \log n , \, (k+1) \log n]\big)$$

of Poincaré arclength $\log n$, where k ranges over all integers, and where f maps I_k isometrically onto I_{k+1} . On the other hand, $G(p(s)) = e^s$ tends to zero as $s \to -\infty$, so any limit point $\hat{z}$ of $p(s)$ as $s \to -\infty$ must belong to the Julia set $J = \partial K$. Using 3.4, given any neighborhood N of $\hat{z}$ we can find a smaller neighborhood N' so that any I_k which intersects N' is contained in N . Since f maps one endpoint of I_k to the other, this shows that $N \cap f(N) \neq \emptyset$ for every neighborhood N , so that $\hat{z}$ must be a fixed point of f . But the set of all limit points must be connected. (See Problem 5-b.) Since f has only finitely many fixed points, this proves that the ray R_t must land at a single fixed point $\gamma(t)$ of the map f .

The corresponding statement for a ray of period p now follows by applying the argument above to the iterate $g = f^{\circ k}$. Together with 18.9, this completes the proof of 18.10 in the periodic case. Finally, if t belongs to $\mathbb{Q}/\mathbb{Z}$ then it must certainly be eventually periodic under multiplication by n , so it follows by 18.1 that the ray R_t lands. Together with 18.12, this completes the proof. □

The proof that at least one periodic ray lands on a repelling or parabolic point will be based on the following ideas. It clearly suffices to consider the special case of a fixed point at the origin. *Thus we assume that* $0 = f(0)$ *is either repelling or parabolic.*

Definition. By a *backward orbit* for f in the open set $\mathbb{C} \smallsetminus \{0\}$ will be meant an infinite sequence $\mathbf{z} = (z_0, z_1, \cdots)$ of points $z_k \in \mathbb{C} \smallsetminus \{0\}$ which satisfy $z_k = f(z_{k+1})$, so that

$$z_0 \leftarrowtail z_1 \leftarrowtail z_2 \leftarrowtail \cdots \ .$$

Let E be the space consisting of all such backwards orbits which converge to zero, $\lim_{k \to \infty} z_k = 0$. To describe a basic neighborhood of such a point $\mathbf{z}$ we choose a neighborhood U_i of each z_i so that, for i sufficiently large, U_i is a neighborhood of zero and is independent of i, and let $N(U_0, U_1, \ldots)$ be the set of all $(z'_0, z'_1, \ldots)$ in E with $z'_i \in U_i$ for all i. Evidently the projection $\pi_k(z_0, z_1, z_2, \ldots) = z_k$ from E to $\mathbb{C} \smallsetminus \{0\}$ is continuous. In fact, since f has no critical points near 0, each connected component of E can be given the structure of a Riemann surface, so that each π_k will be a local conformal isomorphism near $\mathbf{z}$ for k sufficiently large. Define a conformal isomorphism $\mathbf{f} : E \to E$ by the formula

$$\mathbf{f}(z_0, z_1, \ldots) = (f(z_0), f(z_1), \ldots),$$

so that $\pi_k(\mathbf{f}(\mathbf{z})) = f(\pi_k(\mathbf{z}))$.

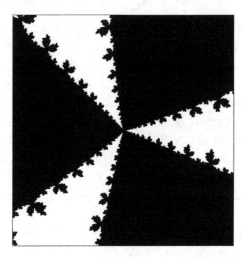

Figure 38. The set $E \cong \mathbb{C} \smallsetminus \{0\}$ for one of the repelling fixed points of Figure 35, with the set $\tilde{K}$ shaded. Here $\mathbf{f}$ permutes the three components of $E \smallsetminus \tilde{K}$, and $\mathbf{f}^{\circ 3}$ maps each component U_k to itself, expanding by a complex factor of $\lambda^3 \simeq 1.36$.

In the repelling case, note that z_k must belong to a linearizing neighborhood of 0 for k sufficiently large. In fact we can define a Kœnigs isomorphism $\kappa : E \xrightarrow{\cong} \mathbb{C} \smallsetminus \{0\}$, with $\kappa(\mathbf{f}(\mathbf{z})) = \lambda\kappa(\mathbf{z})$, by setting

$$\kappa(\mathbf{z}) = \lim_{k \to \infty} \lambda^k z_k$$

where λ is the multiplier. In the parabolic case, replacing f by some iterate if necessary, we can assume that the multiplier is $\lambda = 1$, so that the origin is a fixed point of multiplicity $m \geq 2$. Choose a Leau-Fatou flower, with $m - 1$ repelling petals. For each of these repelling petals P, let E_P be the connected component of E consisting of those $\mathbf{z}$ for which $\{z_k\}$ converges to zero through P. Then the Fatou coordinate

$\phi : P \to \mathbb{C}$ associated with this petal, where $\phi(f(z)) = \phi(z) + 1$ on $P \cap f^{-1}(P)$, gives rise to a conformal Fatou isomorphism $\phi_P : E_P \xrightarrow{\cong} \mathbb{C}$ with $\phi_P(\mathbf{f}(\mathbf{z})) = \phi_P(\mathbf{z}) + 1$, defined by $\phi_P(\mathbf{z}) = \phi(z_k) + k$ for large k.

Let $\tilde{K} \subset E$ be the closed subset consisting of those $\mathbf{z} \in E$ for which the components z_k belong to the filled Julia set K.

Lemma 18.14. *Each connected component U_0 of $E \smallsetminus \tilde{K}$ is a universal covering of $\mathbb{C} \smallsetminus K$ under the projection*

$$\pi_0(z_0, z_1, \ldots) = z_0$$

from U_0 to $\mathbb{C} \smallsetminus K$.

Proof. To show that $\pi_0 : E \smallsetminus \tilde{K} \to \mathbb{C} \smallsetminus K$ is actually a covering map, we start with any simply connected open set $V \subset \mathbb{C} \smallsetminus K$, and any $\mathbf{z} \in E \smallsetminus \tilde{K}$ with $z_0 = \pi_0(\mathbf{z}) \in V$. Since each $f^{\circ k} : \mathbb{C} \smallsetminus K \to \mathbb{C} \smallsetminus K$ is a covering map, there is a unique branch g_k of $f^{-k}|_V$ such that $g_k(z_0) = z_k$. Choose k_0 so that z_k belongs to a linearizing neighborhood or repelling petal for $k \geq k_0$, and choose a smaller neighborhood V' of z_0 so that $g_{k_0}(V')$ is compactly contained in this linearizing neighborhood or petal. Then it follows that the maps $g_k|_{V'}$ converge uniformly to zero. In fact, since the sequence of maps $g_k|_V$ clearly forms a normal family, we obtain the sharper statement that the sequence $g_k|_V$ converges locally uniformly to zero. For otherwise, if the g_k restricted to some compact subset of V had a non-zero limit point, then we could choose a subsequence converging locally uniformly to a non-zero limit. This is impossible, since this limit function must be identically zero on V'. Now it follows that the correspondence

$$z \mapsto \mathbf{g}(z) = (z, g_1(z), g_2(z), \ldots)$$

lifts the neighborhood V to $E \smallsetminus \tilde{K}$, with $\pi_0(\mathbf{g}(z)) = z$. Thus V is evenly covered, as required.

Finally, note that each connected component U_0 of $E \smallsetminus \tilde{K}$ is simply connected. For given any loop $\mathbf{h} : \mathbb{R}/\mathbb{Z} \to U_0$, setting $\mathbf{h}(t) = (h_0(t), h_1(t), \ldots)$ we can choose k so that $h_k(\mathbb{R}/\mathbb{Z})$ is contained in a linearizing neighborhood or petal. The region enclosed by this image loop cannot contain any points of the connected set K, and it follows easily that the loop $\mathbf{h}(\mathbb{R}/\mathbb{Z})$ can be contracted within U_0. Hence the covering map $U_0 \to \mathbb{C} \smallsetminus K$ is a universal covering. $\square$

Main Lemma 18.15. *Each component U_0 of $E \smallsetminus \tilde{K}$ is mapped onto itself by some iterate of $\mathbf{f}$.*

First consider the repelling case, with $\kappa : E \xrightarrow{\cong} \mathbb{C} \smallsetminus \{0\}$. The proof of 18.15 begins as follows. Consider the family of concentric circles $\mathcal{C}_q \subset E$, defined by the equation $|\kappa(\mathbf{z})| = |\lambda|^q$ for $q \geq 0$, where λ is the multiplier so that $\mathbf{f}(\mathcal{C}_q) = \mathcal{C}_{q+1}$.

Lemma 18.16. *Some image $U_k = f^{\circ k}(U_0)$ has the following property. Using the Poincaré metric associated with the simply connected open set U_k, the distance between $U_k \cap \mathcal{C}_0$ and $U_k \cap \mathcal{C}_q$ within U_k is less than or equal to $q \log n$.*

Proof. Note that any two such images U_k and U_ℓ are either equal or disjoint. Let U be the union of these open sets U_k. Consider the Green's function $G \circ \pi_0 : E \to \mathbb{R}$, which is harmonic and strictly positive on U and identically zero on $\partial U \subset \tilde{K}$. Let $G_0 = G \circ \pi_0(\hat{\mathbf{z}})$ be the maximum value which is attained by $G \circ \pi_0$ on the set $U \cap \mathcal{C}_0$. Since $f^{\circ q}(U \cap \mathcal{C}_0) = U \cap \mathcal{C}_q$ and

$$G \circ \pi_0 \circ \mathbf{f}^{\circ q}(\mathbf{z}) \;=\; n^q\, G \circ \pi_0(\mathbf{z}) \,,$$

it follows that the maximum value of $G \circ \pi_0$ on $U \cap \mathcal{C}_q$ is $n^q G_0$, attained at the point $\mathbf{f}^{\circ q}(\hat{\mathbf{z}})$. In U_k, just as in $\mathbb{C} \smallsetminus K$, the orthogonal trajectories of the equipotentials

$$G \circ \pi_0(\mathbf{z}) = \text{constant}$$

can be described as external rays, and can be parametrized by their Poincaré arclength $\log G \circ \pi_0$. If we start at the point $\hat{\mathbf{z}}$, which belongs to $U_k \cap \mathcal{C}_0$ for some k, and follow the external ray through this point until we first arrive at some point $\mathbf{z}' \in U_k \cap \mathcal{C}_q$, then this ray segment will have Poincaré length $\log G \circ \pi_0(\mathbf{z}') - \log G_0 \leq \log n^q = q \log n$, as required. $\square$

On the other hand, suppose that 18.15 were false, so that the sets U_k were pairwise disjoint. Then we will prove that the Poincaré distance between $U_k \cap \mathcal{C}_0$ and $U_k \cap \mathcal{C}_q$ within U_k would have to increase more than linearly with q. This will contradict 18.16, and hence prove 18.15. The argument will be based on the following very rough estimate.

Lemma 18.17. *Consider a strip $S \subset \mathbb{C}$ of width w, bounded by two parallel lines L_1 and L_2. Let U be a simply connected region which intersects both boundary lines, and let A be the Euclidean area of $U \cap S$. Then the Poincaré distance between $U \cap L_1$ and $U \cap L_2$ within U is greater than $w\sqrt{\frac{\pi}{4A}} - 1$.*

In particular, as A tends to zero with w fixed, this distance tends to infinity. The proof will be based the following much more precise inequality: *For any path segment γ in U, the Poincaré arclength of γ is greater than or equal to*

$$\frac{1}{2} \int_\gamma \frac{|dz|}{\text{dist}(z, \partial U)} \,,$$

using the Euclidean distance to the boundary. For a proof, see A.8 in Appendix A. Using this inequality, the proof of 18.17 proceeds as follows. Define r by the equation $A = \pi r^2$, and let z be any point of $U \cap S$ which has distance at least r from the lines L_i . Since the closed disk of radius r about z has area equal to $A = \mathrm{area}(U \cap S)$, it must intersect the boundary of U . Therefore the Poincaré metric at z is at least $|dz|/2r$. Choose a minimal Poincaré geodesic γ within U between the two boundary lines of S . Then that part of γ which has distance $> r$ from ∂S must have Euclidean length at least $w - 2r$, and hence Poincaré length at least

$$\frac{w - 2r}{2r} = w\sqrt{\frac{\pi}{4A}} - 1 \, ,$$

as required.   □

We will make use of the flat metric $|d\kappa|/\kappa$ on the set $E \cong \mathbb{C} \smallsetminus \{0\}$. (Equivalently, we could identify E with the bi-infinite cylinder $\mathbb{C}/(2\pi i \mathbb{Z})$ with its usual flat Euclidean metric, using the conformal equivalence $\mathbf{z} \mapsto \log \kappa(\mathbf{z})$.) Using this metric, note that $\mathbf{f}$ maps E isometrically onto itself. Let $A_q \subset E$ be the annulus bounded by the circles $\mathcal{C}_q$ and $\mathcal{C}_{q+1}$, and let α stand for area with respect to this flat metric. Assuming that the U_k are pairwise disjoint, since $\alpha(A_0)$ is finite it follows that $\alpha(U_k \cap A_0)$ tends to zero as $|k| \to \infty$. Since $\mathbf{f}^{\circ q}$ maps the intersection $U_{-q} \cap A_0$ isometrically onto $U_0 \cap A_q$, it follows that $\alpha(U_0 \cap A_q)$ also tends to zero as $q \to \infty$. Using 18.17, it follows that the Poincaré distance, within U_0 , between $\mathcal{C}_q$ and $\mathcal{C}_{q+1}$ tends to infinity as $q \to \infty$. Hence the Poincaré distance within U_0 between $\mathcal{C}_0$ and $\mathcal{C}_q$ increases more than linearly with q . A similar argument applies to the Poincaré distance within each U_k . But this contradicts 18.16, and this contradiction completes the proof of 18.15 in the repelling case.

The proof of 18.11 in the repelling case now proceeds as follows. By 18.15 we can choose $k > 0$ so that $\mathbf{f}^{\circ k}$ maps the simply connected set U_0 biholomorphically onto itself, evidently without fixed points. Hence we can form a quotient Riemann surface $\mathcal{S}$ by identifying each $\mathbf{z} \in U_0$ with $\mathbf{f}^{\circ k}(\mathbf{z})$. We know from §2 that $\mathcal{S}$ can only be an annulus or a punctured disk. However, the punctured disk case is impossible, since 18.17 easily yields a positive lower bound for the Poincaré arclength of a path joining $\mathbf{z}$ to $\mathbf{f}^{\circ k}(\mathbf{z})$. Thus $\mathcal{S}$ is an annulus. In particular, there is a unique simple closed Poincaré geodesic on $\mathcal{S}$. (Compare Problem 2-f. Intuitively, think of a rubber band placed around a napkin ring which shrinks to the unique simple closed curve of minimal length.) Lifting to the universal covering U_0 of $\mathcal{S}$, we obtain an $\mathbf{f}^{\circ k}$-invariant bi-infinite Poincaré geodesic.

Now projecting to $\mathbb{C} \smallsetminus K$ we obtain an $f^{\circ k}$-invariant bi-infinite Poincaré geodesic $p : \mathbb{R} \to \mathbb{C} \smallsetminus K$. Since the Green's function $G(p(s))$ tends to infinity as $s \to +\infty$, this can only be an external ray. (Any Poincaré geodesic in the right half-plane is either horizontal, corresponding to an external ray, or else has bounded real part.) On the other hand, since the Kœnigs coordinate of $p(s)$ tends to zero as $s \to -\infty$, this ray lands at the origin, as required. This proves 18.11 in the repelling case. □

Proofs in the parabolic case. As noted earlier, it suffices to consider the case $\lambda = 1$. Recall that the set E_P consists of backward orbits which converge to zero through a given repelling petal P, with Fatou isomorphism $\phi_P : E_P \overset{\cong}{\longrightarrow} \mathbb{C}$ satisfying $\phi_P \circ \mathbf{f}(\mathbf{z}) = \phi_P(\mathbf{z}) + 1$. If $\phi_P(\mathbf{z}) = u + iv$ with $|v|$ sufficiently large and u close to $-\infty$, note that $\pi(\mathbf{z}) \in \mathbb{C}$ must belong to one of the two adjacent attracting petals, and hence must belong to the filled Julia set K. It follows that the entire set $\phi_P(E_P \smallsetminus \tilde{K}) \subset \mathbb{C}$ must be contained in a strip $|v| <$ constant, of finite height. In place of the circles $\mathcal{C}_q$ and annuli A_q of the argument above, we now use the vertical lines

$$L_q = \{\mathbf{z} \in E_P ; \ \operatorname{Re}(\phi_P(\mathbf{z})) = q\}$$

and the vertical bands bounded by L_q and L_{q+1}. An argument completely analogous to that of 18.16 shows that the Poincaré distance between L_0 and L_q within a suitable U_k is less than or equal to $q \log n$. On the other hand, if the images $U_k = \mathbf{f}^{\circ k}(U_0)$ were pairwise disjoint, then an area argument, just like that above, would show that this distance must grow more than linearly with q. This contradiction completes the proof of 18.15.

In fact (in this parabolic case with $\lambda = 1$), we get the sharper statement that $\mathbf{f}(U_0) = U_0$. For if $U_1 = \mathbf{f}(U_0) \neq U_0$, then the image $\phi_P(U_1)$ in the strip $\{u + iv ; \ |v| < \text{constant}\}$ would have to lie either above or below the image $\phi_P(U_0)$. If for example it were above, then clearly $\phi_P(U_2)$ would be above $\phi_P(U_1)$, and so on, so that we could not have $U_k = U_0$ for any $k > 0$.

Now, just as in the argument above, we can form a quotient annulus $\mathcal{S}$ by identifying each $\mathbf{z} \in U_0$ with $\mathbf{f}(\mathbf{z})$. The unique simple closed geodesic in $\mathcal{S}$ lifts to a bi-infinite geodesic in U_0, which projects to an external ray $R_t = f(R_t)$ which land at the origin through the given repelling petal P. This completes the proof of 18.11 and 18.13. □

Further development and application of these ideas may be found for example in Goldberg and Milnor, in Kiwi 1995, 1997, in Schleicher 1997, and in Milnor 1999b.

§19. Hyperbolic and Subhyperbolic Maps

This section will describe some examples of locally connected Julia sets, using arguments due to Sullivan, Thurston, Douady and Hubbard.

Definition. A rational map f will be called dynamically *hyperbolic* if f is *expanding* on its Julia set J in the following sense: There exists a Riemannian metric μ, defined on some neighborhood of J, such that the derivative Df_z at every point $z \in J$ satisfies the inequality

$$\|Df_z(v)\|_\mu > \|v\|_\mu$$

for every non-zero vector v in the tangent space $T\widehat{\mathbb{C}}_z$. (Notation as in 2.11.) Since J is compact, it follows that there exists an *expansion constant* $k > 1$ with the property that $\|Df_z\|_\mu \geq k$ for all points z in some neighborhood of J. In particular, any smooth path of length L in this neighborhood maps to a smooth path of length $\geq kL$. It follows easily that every $z \in J$ has some open neighborhood N_z in $\widehat{\mathbb{C}}$ such that the associated Riemannian distance function satisfies

$$\operatorname{dist}_\mu(f(x), f(y)) \geq k \cdot \operatorname{dist}_\mu(x, y) \qquad (19:1)$$

for all $x, y \in N_z$.

Recall from 11.17 that the *postcritical set* P of f is the collection of all forward images $f^{\circ j}(c)$ with $j > 0$, where c ranges over the critical points of f.

Theorem 19.1. Hyperbolic Maps. *A rational map of degree $d \geq 2$ is dynamically hyperbolic if and only if its postcritical closure $\overline{P}$ is disjoint from its Julia set, or if and only if the orbit of every critical point converges to an attracting periodic orbit. In fact if f is hyperbolic, then every orbit in its Fatou set must converge to an attracting periodic orbit.*

Remark. Hyperbolic maps have many other important properties. Every periodic orbit for a hyperbolic map must be attracting or repelling. If f is hyperbolic, then every nearby map is also hyperbolic. Furthermore, according to Mañé, Sad, Sullivan (and also Lyubich 1983, 1990), the Julia set $J(f)$ deforms continuously under a deformation of f through hyperbolic maps. (In contrast, in the non-hyperbolic case, a small change in f may well lead to a drastic alteration of $J(f)$.) The well known but unproved "Generic Hyperbolicity Conjecture" is the claim that every rational map can be approximated arbitrarily closely by a hyperbolic map.

Proof of 19.1. Let V be the complement $\widehat{\mathbb{C}} \smallsetminus \overline{P}$, and let $W = f^{-1}(V) \subset \widehat{\mathbb{C}}$. As in 11.17, we see that $W \subset V$, and that f maps W onto

V by a d-fold covering map. Furthermore, if we exclude the trivial case of a map which is conjugate to $z \mapsto z^{\pm d}$, then every connected component of V or W is conformally hyperbolic.

Now suppose that $\overline{P} \cap J = \emptyset$, or in other words that $J \subset V$. Then W must be strictly smaller than V. For otherwise V would map to itself under f, and hence be contained in the Fatou set. In fact any connected component of W which intersects J must be strictly smaller than the corresponding component of V. It follows, arguing as in (11 : 6), that $\|Df_z\|_V > 1$ for every $z \in W$. (Here the subscript V indicates that we use the Poincaré metric associated with the hyperbolic surface V.) Since $J \subset W$, this proves that f is dynamically hyperbolic.

(Remark: An alternative version of this argument would be based on the observation that f^{-1} must lift to a single valued map F from the universal covering surface $\tilde{V}$ to itself. Then F must be distance decreasing for the Poincaré metric on $\tilde{V}$, hence f must be distance increasing for the Poincaré metric on V. Compare the proof of 19.6.)

Conversely, suppose that f is dynamically hyperbolic. Thus we can choose some Riemannian metric on a neighborhood V of J so that f is expanding with expansion factor $\geq k > 1$ throughout some smaller neighborhood V' of J. It certainly follows that there cannot be any critical point in V'. Choose $\epsilon > 0$ small enough so that

(1) every point in the open ϵ-neighborhood $N_\epsilon(J)$ can be joined to J within V by at least one minimal geodesic; and so that

(2) $$f^{-1} N_\epsilon(J) \subset V'.$$

For any $z \in f^{-1} N_\epsilon(J) \smallsetminus J$, it follows that

$$\text{dist}(z, J) \leq \text{dist}(f(z), J)/k < \text{dist}(f(z), J).$$

In fact, if we choose a minimal geodesic from $f(z)$ to J within V, then one of its d preimages will join the point z to J, and will have length at most $\text{dist}(f(z), J)/k$.

It follows that an arbitrary orbit $z_0 \mapsto z_1 \mapsto \cdots$ in the Fatou set can contain at most a finite number of points in $N_\epsilon(J)$. In fact, no point outside of $N_\epsilon(J)$ can map into $N_\epsilon(J)$, while if an orbit starts in $N_\epsilon(J) \smallsetminus J$, then the distance between z_i and J must increase by a factor of k or more with each iteration until the orbit leaves $N_\epsilon(J)$, never to return. Therefore any accumulation point $\hat{z}$ for this orbit lies in the Fatou set. If U is the Fatou component containing $\hat{z}$, then evidently some iterate $f^{\circ p}$ must map U into itself. According to the classification of periodic Fatou components in 16.1, U must be either an attracting basin, a parabolic basin, or a rotation domain. Since U clearly cannot be a parabolic basin, and by 11.17 and 15.7 cannot be a rotation domain, it must be an attracting basin. Therefore the orbit $z_0 \mapsto z_1 \mapsto \cdots$ must converge to the associated attracting periodic orbit. In particular, the orbit of any critical point must converge to an attracting periodic orbit. This clearly implies that $\overline{P} \cap J = \emptyset$, and completes the proof of 19.1. $\square$

Theorem 19.2. Local Connectivity. *If the Julia set of a hyperbolic map is connected, then it is locally connected.*

The proof will be based on three lemmas.

Lemma 19.3. Fatou Component Boundaries. *If U is a simply connected Fatou component for a hyperbolic map, then the boundary ∂U is locally connected.*

Remark. In the special case of a polynomial map, note that this result by itself implies 19.2. For the basin of infinity for a polynomial map is an invariant Fatou component, with boundary equal to the entire Julia set.

Proof of 19.3. First consider the case of an invariant component $U = f(U)$. Choose a conformal isomorphism $\phi : \mathbb{D} \xrightarrow{\cong} U$ so that $\phi(0)$ is the attracting fixed point in U. Then $F = \phi^{-1} \circ f \circ \phi$ is a proper holomorphic map from $\mathbb{D}$ to itself, with $F(0) = 0$, and with at least one critical point by 8.6. For any $0 < r < 1$, it follows from the Schwarz Lemma that F maps the disk $\mathbb{D}_r$ of radius r into some disk of strictly smaller radius.

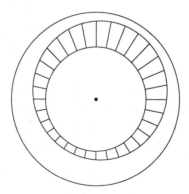

If r is sufficiently close to one, then it is not hard to see that the closure of the annulus $A_0 = F^{-1}(\mathbb{D}_r) \smallsetminus \overline{\mathbb{D}}_r$ is fibered by radial line segments (as illustrated above). That is, each radial line $\{te^{i\theta} ; 0 < t < 1\}$ intersects $\overline{A}_0$ in a closed interval I_θ which varies smoothly with θ. To see this, express F as a Blaschke product

$$F(w) = e^{i\alpha} \prod (w - a_j)/(1 - \overline{a}_j w) .$$

If $w\bar{w} = 1$, then a brief computation shows that the radial derivative

$$\frac{d\log F(w)}{d\log(w)} = w\frac{F'(w)}{F(w)}$$

takes the form $\sum(1 - a_j\bar{a}_j)/|w - a_j|^2 > 0$. Hence for $|w|$ close to $+1$ the real part of this radial derivative is still positive, which implies transversality of the required intersection.

Since f is hyperbolic, we can choose a Riemannian metric in some neighborhood of $J \supset \partial U$ so that f is expanding near J with expansion constant $k > 1$. Choose $r < 1$ large enough so that $\phi(\mathbb{D} \smallsetminus \mathbb{D}_r)$ is contained in this neighborhood. Using the induced metric on $\mathbb{D} \smallsetminus \mathbb{D}_r$, let M be the maximum of the lengths of the radial intervals I_θ. Now consider the sequence of annuli $A_0, A_1, A_2, \ldots$, converging to the boundary of $\mathbb{D}$, where $A_m = F^{-m}(A_0)$. Note that the closure of each A_m is fibered by the connected components of the preimages $F^{-m}(I_\theta)$, and that each such component curve segment has length at most M/k^m. Hence we can inductively construct a sequence of homeomorphisms

$$g_m : \partial\mathbb{D}_r \mapsto F^{-m}\partial\mathbb{D}_r$$

so that g_0 is the identity map, and so that $g_m(re^{i\theta})$ and $g_{m+1}(re^{i\theta})$ are the two endpoints of the same fiber in the annulus A_m. Then

$$\text{dist}(g_m(re^{i\theta}), g_{m+1}(re^{i\theta})) \leq M/k^m,$$

hence the maps

$$\phi \circ g_m : \partial\mathbb{D}_r \to U \subset \hat{\mathbb{C}}$$

form a Cauchy sequence. Therefore they converge uniformly to a continuous limit map from $\partial\mathbb{D}_r$ onto ∂U. By 17.14, this proves that ∂U is locally connected.

The case of a periodic Fatou component $U = f^{\circ p}(U)$ now follows by applying this argument to the iterate $f^{\circ p}$. Since any Fatou component U is eventually periodic by 19.1 (or by Sullivan's Theorem), and since ∂U is locally homeomorphic to $\partial f^{\circ q}(U)$, the conclusion follows. $\square$

For the next lemma, we use the spherical metric on $\hat{\mathbb{C}}$.

Lemma 19.4. Most Fatou Components are Small. *If f is hyperbolic with connected Julia set, then for every $\epsilon > 0$ there are only finitely many Fatou components with diameter greater than ϵ.*

In other words, if there are infinitely many Fatou components, numbered in any order as $U_1, U_2, \ldots$, then the diameter of U_j in the spherical metric must tend to zero as $j \to \infty$. (However a hyperbolic map with disconnected Julia set may well have infinitely many Fatou components with diameter with diameter bounded away from zero. See Figure 2a for McMullen's example. I don't know any such example with connected Julia set, even in the non-hyperbolic case.)

Proof of 19.4. Since f is hyperbolic, we can choose some Riemannian metric μ on a neighborhood V of the Julia set J so that f is expanding with respect to μ on some neighborhood N of J, with $\overline{N} \subset V$. For any Fatou component U_j which is contained in N, define the distance $\mathrm{dist}_{\mu|U_j}(x, y)$ between two points of U_j to be the infimum of the μ-lengths of paths between x and y, where we only allow paths which are contained in U_j. Then we can define the diameter $\mathrm{diam}_{\mu|U_j}(U_j)$ to be the supremum of $\mathrm{dist}_{\mu|U_j}(x, y)$ for $x, y \in U_j$. It follows easily from the proof of 19.3 that this diameter is always finite. Note that the various Fatou components U_j, together with N, form an open cover of the compact space $\widehat{\mathbb{C}}$. Choosing a finite subcover, we see that all but finitely many of the U_j are contained in N.

For each U_j, we know from 19.1 that some forward image $f^{\circ \ell}(U_j)$ contains an attracting periodic point, and hence is not contained in N. Define the *level* of U_j to be the smallest $\ell \geq 0$ such that $f^{\circ \ell}(U_j) \not\subset N$. Since there are only finitely many U_j of level zero, it follows that there are only finitely many of each fixed level. Let M be the maximum of $\mathrm{diam}_{\mu|U_j}(U_j)$ among the U_j of level one. If $k > 1$ is the expansion constant, it follows that

$$\mathrm{diam}_{\mu|U_j}(U_j) \leq M/k^{\ell-1}$$

for each U_j of level $\ell \geq 1$. Evidently this tends to zero as $\ell \to \infty$. Since all but finitely many of the U_j lie within the compact set $\overline{N}$, it follows easily that the diameters in the spherical metric also tend to zero, as required. $\square$

Lemma 19.5. Characterization of Locally Connected Sets in S^2. *If X is a compact subset of the sphere S^2 such that every component of $S^2 \smallsetminus X$ has locally connected boundary, and such that there are at most finitely many components with diameter $> \epsilon$ for any $\epsilon > 0$, then X is locally connected.*

(For the converse statement, see Problem 19-f.)

Proof. Given any neighborhood $N_\epsilon(x)$ of radius ϵ, using the spherical metric, for a point $x \in X$, we will find a smaller neighborhood $N_\delta(x)$ so that any point in $X \cap N_\delta(x)$ is joined to x by a connected subset of $X \cap N_\epsilon(x)$. In fact, choose $\delta < \epsilon/2$ so that, for any component U_j of

$S^2 \smallsetminus X$ with diameter $\geq \epsilon/2$, any two points of ∂U_j with distance less than δ are joined by a connected subset of ∂U_j of diameter less than $\epsilon/2$. (See 17.13.) Now for any $y \in X \cap N_\delta(x)$, take the geodesic I from x to y and replace each connected component of $I \smallsetminus X$ by a connected subset of the boundary of the corresponding U_j with diameter less than $\epsilon/2$. The closure of the result will be a connected subset of $X \cap N_\epsilon$ containing both x and y. This proves that X is locally connected at the arbitrary point x. $\square$

Clearly Theorem 19.2 follows immediately from these three lemmas. $\square$

Douady and Hubbard, using ideas of Thurston, also consider a wider class of mapping which they call *subhyperbolic*. These may have critical points in the Julia set, but only if their orbits are eventually periodic. The only change in the definition is that the Riemannian metric in a neighborhood of J is now allowed to have a finite number of relatively mild singularities in the postcritical set. To understand the allowed singularities, consider a smooth conformal metric $\rho(w)|dw|$ which is invariant under rotation of the w-plane through an angle of $2\pi/m$ radians, and consider the identification space in which w is identified with $e^{2\pi i/m}w$. The resulting object is a smooth Riemannian manifold except at the origin, where it has a "cone point". If we set $z = w^m$, then the induced metric in the z-plane has the form $\gamma(z)|dz|$ where

$$\gamma(z) \;=\; \rho(w)\left|\frac{dz}{dw}\right|^{-1} \;=\; \frac{\rho(\sqrt[m]{z})}{m\,|z|^{1-1/m}} \;.$$

Thus $\gamma(z) \to \infty$ as $z \to 0$, but the singularity is relatively innocuous since any reasonable path $t \mapsto z(t)$ still has finite length $\int \gamma(z(t))|dz(t)|$.

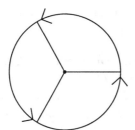

Figure 39. Disk in the w-plane on the left with three fundamental domains under $120°$ rotation indicated, and the quotient space in a pushed forward metric on the right.

Definition. A conformal metric on a Riemann surface, with the expression $\gamma(z)|dz|$ in terms of a local uniformizing parameter z, will be called an *orbifold metric* if the function $\gamma(z)$ is smooth and non-zero except at a locally finite collection of "ramified points" $a_1, a_2, \ldots$ where it blows up in the following special way. There should be integers $\nu_j \geq 2$ called the *ramification indices* at the points a_j such that, if we take a local branched covering by setting $z(w) = a_j + w^{\nu_j}$, then the induced metric $\gamma(z(w))|dz/dw| \cdot |dw|$ on the w-plane is smooth and non-singular throughout some neighborhood of the origin. We will say that f is *expanding* with repect to such a metric if its derivative satisfies $\|Df_p\| \geq k > 1$ whenever p and $f(p)$ are not ramified points. (Note that we cannot expect the sharper condition (19 : 1) near a critical point.)

Definition. The rational map f is *subhyperbolic* if it is expanding with respect to some orbifold metric on a neighborhood of its Julia set.

Following Douady and Hubbard, we have the following two results. (Compare 14.5.)

> **Theorem 19.6.** *A rational map is subhyperbolic if and only if every critical orbit is either finite, or converges to an attracting periodic orbit.*

> **Theorem 19.7.** *If f is subhyperbolic with $J(f)$ connected, then $J(f)$ is locally connected.*

The proof of 19.7 is essentially identical to the proof of 19.2. In particular, the proofs of 19.3 and 19.4 work equally well in the subhyperbolic case. Details will be left to the reader. □

Proof of 19.6. The argument is just an elaboration of the proof of 19.1. In one direction, if f is expanding with respect to some orbifold metric defined near the Julia set, then just as in 19.1 there is a neighborhood $N_\epsilon(J)$ so that every orbit in the Fatou set eventually leaves this neighborhood, never to return. Hence it can only converge to an attractive periodic orbit. On the other hand, if c is a critical point in the Julia set, then every forward image $f^{\circ i}(c)$, $i > 0$, must be one of the ramification points a_j for our orbifold metric, since the map $f^{\circ i}$ has derivative zero at the critical point c, and yet must satisfy $\|Df_z^{\circ i}\| \geq k^i$ at points arbitrarily close to c. The collection of ramified points in J is required to be locally finite, so it follows that the orbit of c must be eventually periodic.

For the proof in the other direction, we must introduce more ideas from Thurston's theory of orbifolds. (See Appendix E for a brief introduction to this theory.)

Definition. For our purposes, an *orbifold* (S, ν) will just mean a Riemann surface S, together with a locally finite collection of marked points a_j (to be called ramified points), each of which is assigned a *ramification* *index* $\nu_j = \nu(a_j) \geq 2$ as above. For any point z which is not one of the a_j we set $\nu(z) = 1$.

To every rational map f which satisfies the conditions of Theorem 19.6, we assign the *canonical orbifold* (S, ν) as follows. As underlying Riemann surface S we take the Riemann sphere $\widehat{\mathbb{C}}$ with all attracting periodic orbits removed. As ramified points a_j we take all (strictly) *postcritical* points, that is, all points which have the form $a_j = f^{\circ i}(c)$ for some $i > 0$, where c is a critical point of f. Since every critical orbit is either finite or converges to a periodic attractor, we see easily that this collection of points a_j is locally finite in S (although perhaps not in $\widehat{\mathbb{C}}$). In order to specify the corresponding ramification indices $\nu_j = \nu(a_j)$, we will need another definition. If $f(z_1) = z_2$, with local power series development

$$f(z) = z_2 + b(z - z_1)^n + \text{(higher terms)},$$

where $b \neq 0$ and $n \geq 1$, then the integer $n = n(f, z_1)$ is called the *local degree* or the *branch index* of f at z_1. *Choose the* $\nu(a_j) \geq 2$ *to be the smallest integers which satisfy the following:*

> **Condition** $(\star)$. For any $z \in S$, the ramification index $\nu(f(z))$ at the image point must be some multiple of the product $n(f, z) \nu(z)$.

(Note: If $f(z)$ is an attracting periodic point, and hence not in S, then we set $\nu(f(z)) = \infty$.) To construct these integers $\nu(a_j)$, where a_j ranges over all postcritical points in S, we consider all pairs (c, m) where c is a critical point with $f^{\circ m}(c) = a_j$, and choose $\nu(a_j)$ to be the least common multiple of the corresponding branch indices $n(f^{\circ m}, c)$. (Note that a_j itself may be a critical point, since one critical point may eventually map to another.) There are only finitely many such pairs (c, m) since we have removed all superattracting periodic orbits, so this least common multiple is well defined and finite. It is not hard to check that it provides a minimal solution to the required Condition $(\star)$.

As in Appendix E, we consider the *universal covering surface*

$$\tilde{S}_\nu \to (S, \nu)$$

for this canonical orbifold, that is the unique regular branched covering of S which is simply connected and has the given $\nu : S \to \mathbb{Z}$ as ramification function. Such a universal covering could fail to exist only if S were the entire Riemann sphere with at most two ramified points (see E.1 in the appendix), and it is straighforward to show that this case can never occur for our canonical orbifold.

Since $f^{-1}(S) \subset S$, it is not difficult to see that f^{-1} lifts to a single valued holomorphic map $F : \tilde{S}_\nu \to \tilde{S}_\nu$. In fact Condition $(\star)$ is exactly what is needed in order to guarantee that such a lifting of f^{-1} exists locally, and since $\tilde{S}_\nu$ is simply connected there is no obstruction to extending to a global lifting.

Conformally hyperbolic case. If $\tilde{S}_\nu$ is hyperbolic, then F must either preserve or decrease the Poincaré metric for this universal covering surface. If F were metric preserving, then f would preserve the orbifold metric for (S, ν), and it would follow that every periodic point for f in S must be indifferent, which is impossible. Hence F must be metric decreasing, and f must be metric increasing. Since J is compact, it follows that $\|DF_w\| \le 1/k < 1$ whenever $w \in \tilde{S}_\nu$ projects into a suitably chosen neighborhood W of J. Hence $\|Df_z\| \ge k > 1$ for every $z \in W$ such that z and $f(z)$ are not ramified points.

Conformally Euclidean case. (Compare 19.9.) The easiest way to proceed when $\tilde{S}_\nu$ is a Euclidean surface is simply to change the ramification function ν. For example if we choose some periodic orbit in S, and replace ν by a ramification function ν' which is equal to 2ν on this orbit and equal to ν elsewhere, then Condition $(\star)$ will be preserved while $\tilde{S}_{\nu'}$, as a non-trivial branched covering of $\tilde{S}_\nu \cong \mathbb{C}$, will certainly be hyperbolic. The proof then goes through as above.

Spherical case. If $\tilde{S}_\nu$ were conformally equivalent to $\hat{\mathbb{C}}$, then S would have to be the whole Riemann sphere. Furthermore, the composition

$$\tilde{S}_\nu \xrightarrow{F} \tilde{S}_\nu \xrightarrow{\text{projection}} S \xrightarrow{f} S$$

would have to coincide with the projection map, and yet its degree would have to be strictly larger than the degree of the projection map. Thus this case can never occur. This completes the proof of 19.6. □

As a corollary, we obtain another proof of 16.5.

Corollary 19.8. *If f is subhyperbolic with no attracting periodic orbits, so that S is the entire Riemann sphere, then f is expanding with respect to its orbifold metric on the entire sphere. Hence the Fatou set is vacuous, and $J(f)$ is the entire sphere.*

In the Euclidean case, a more careful argument yields a much more precise description of the subhyperbolic map f. In order to determine the geometry of $\tilde{S}_\nu$, we introduce the *orbifold Euler characteristic*

$$\chi(S, \nu) = \chi(S) + \sum \left(\frac{1}{\nu(a_j)} - 1 \right),$$

to be summed over all points a_j with $\nu(a_j) \ne 1$. (Here $\chi(S)$ is the ordinary Euler characteristic, equal to $2 - m$ where m is the number of points in the complement $\widehat{\mathbb{C}} \smallsetminus S$.) It follows easily from E.4 that the universal covering surface $\tilde{S}_\nu$ is either spherical, hyperbolic, or Euclidean according as $\chi(S, \nu)$ is positive, negative, or zero.

Theorem 19.9. *If $\chi(S, \nu) = 0$, then f induces a linear isomorphism $\tilde{f}(w) = \alpha w + \beta$ from the Euclidean covering space $\tilde{S}_\nu \cong \mathbb{C}$ onto itself. In this case, the Julia set is either a circle or line segment, or the entire Riemann sphere. Here the expansion constant $|\alpha|$ is equal to the degree d when J is one-dimensional, and is equal to $\sqrt{d}$ when J is the entire sphere.*

Compare §7. (Caution: The coefficient α itself is not uniquely determined, since the lifting of f to the covering surface is determined only up to composition with a deck transformation. The deck transformations may well have fixed points, since we are dealing with a branched covering, but necessarily have the form $w \mapsto \alpha' w + \beta'$ where α' is some root of unity.)

Proof of 19.9. Starting with the orbifold (S, ν) associated with a subhyperbolic map f, construct a new orbifold (S', μ) as follows. Let $S' = f^{-1}(S)$ be the open set S with all immediate pre-images of attracting periodic points removed, and define $\mu = f^*(\nu)$ by the formula

$$\mu(z) = \nu(f(z))/n(f, z)$$

where $n(f, z)$ is the branch index. In follows from Condition $(\star)$ that $\mu(z)$ is an integer with $\mu(z) \geq \nu(z)$ for all $z \in S'$. Evidently it follows that

$$\chi(S', \mu) \leq \chi(S, \nu)$$

with equality only if $S' = S$ and $\mu = \nu$. But by Lemma E.2 in the appendix, since the map $f : (S', \mu) \to (S, \nu)$ is a "d-fold covering of orbifolds", we conclude that f induces an isomorphism $\tilde{S}'_\mu \to \tilde{S}_\nu$ of universal covering surfaces, and also that the Riemann-Hurwitz formula takes the form

$$\chi(S', \mu) = \chi(S, \nu)\,d.$$

Combining these two statements, we see that

$$\chi(S, \nu)\,d \leq \chi(S, \nu),$$

with equality if and only if $S = f^{-1}(S)$ and $\nu = f^*\nu$. Since $d \geq 2$, this provides another proof that $\chi(S, \nu) \leq 0$. Furthermore, it shows that we are in the Euclidean case $\chi(S, \nu) = 0$ if and only if $S = f^{-1}(S)$ and $\nu = f^*\nu$, so that f maps (S, ν) to itself by a d-fold covering of orbifolds,

Thus, when $\tilde{S}_\nu$ is conformally Euclidean, it follows that f lifts to a necessarily linear automorphism $\tilde{f}(w) = \alpha w + \beta$ of the universal covering surface $\tilde{S}_\nu \cong \mathbb{C}$. Furthermore, since S is fully invariant under f, it follows from 4.6 that the complement $\widehat{\mathbb{C}} \smallsetminus S$ has at most two points. We now consider the three possibilities.

Case 0. If $S = \widehat{\mathbb{C}}$, then we can compute the degree by integrating $\|Df_z\|$ over the sphere. In fact, using the (locally Euclidean) orbifold metric, note that f maps a generic small region of area A to a region of area $|\alpha|^2 A$. Integrating over S, we see that the degree d must be equal to $|\alpha|^2$.

Case 1. If $S \cong \mathbb{C}$, then solving the required equation

$$\chi(S, \nu) \;=\; 1 - \sum \left(\frac{1}{\nu(z)} - 1 \right) \;=\; 0 \,,$$

it is not difficult to check that there must be exactly two ramification points, both with index $\nu(z) = 2$. The corresponding universal covering space is isomorphic to $\mathbb{C}$, with the integers as branch points, and with all linear maps of the form $w \mapsto \pm w + m$ as deck transformations. Since $\tilde{f}$ must carry integers to integers, it follows easily that f is a Chebyshev map up to sign, with an interval as Julia set, and with degree $|\alpha|$.

Case 2. If $S \cong \mathbb{C} \smallsetminus \{0\}$, then there can be no ramification points, and it follows that f is conjugate to $z \mapsto z^{\pm d}$. Thus the Julia set is a circle, and again $d = |\alpha|$. $\square$

Other Locally Connected Julia Sets. Many other Julia sets are known to be locally connected. The best known result is the Yoccoz proof that a quadratic polynomial Julia set is locally connected provided that it is connected, with no Cremer points or Siegel disks, and is not infinitely renormalizable. (Compare [Hubbard 1993], [Milnor 1992c].) A rational map is called *geometrically finite* if the orbit of every critical point in its Julia set is eventually periodic. Recall from §16 that every critical orbit in the Fatou set must converge to an attracting cycle or to a parabolic cycle. Thus, in the geometrically finite case we have a very strict control of all critical orbits. In particular, it follows from 14.4 and 15.7 that a geometrically finite map can have no Cremer points, Siegel disks, or Herman rings. [Tan Lei, Yin] and [Pilgrim, Tan Lei] have proved the following much sharper version of 19.6: *If f is geometrically finite, then every connected component of its Julia set is locally connected.* Another important example is given by Petersen's proof for quadratic Siegel disks of bounded type. (Compare [Petersen 1996], as well as [Yampolsky 1995].)

Concluding Problems:

Problem 19-a. The non-wandering set. By definition, the *non-wandering set* for a continuous map $f : X \to X$ is the closed subset $\Omega \subset X$ consisting of all $x \in X$ such that for every neighborhood U of x there exists an integer $k > 0$ such that $U \cap f^{\circ k}(U) \neq \emptyset$. Using the results of §4 and §16, show that the non-wandering set for a rational map f is the (disjoint) union of its Julia set, its rotation domains (if any), and its set of attracting periodic points.

Problem 19-b. Axiom A. In the literature on smooth dynamical systems a one-dimensional* map is said to satisfy Smale's Axiom A if and only if

(1) the non-wandering set Ω splits as the union of a closed subset Ω^+ on which f is infinitesimally expanding with respect to a suitable Riemannian metric, and a subset Ω^- on which f is contracting, and

(2) periodic points are everywhere dense in Ω.

Show that a rational map is hyperbolic if and only if it satisfies Axiom A.

Problem 19-c. An orbifold example. Show that the Julia set for the rational map $z \mapsto (1-2/z)^{2n}$ is the entire Riemann sphere. For $n > 1$, show that the orbifold metric for this example is hyperbolic. For $n = 1$, show that it is Euclidean, but is not a Lattès example.

Problem 19-d. The Euclidean case. For any subhyperbolic map whose canonical orbifold metric is Euclidean, show that every periodic orbit outside of the finite postcritical set has multiplier λ satisfying $|\lambda| = n^{p/\delta}$ where n is the degree, p is the period, and δ is the dimension (1 or 2) of the Julia set.

Problem 19-e. Expansive maps. A rational map f is said to be *expansive* in a neighborhood of its Julia set if there exists $\epsilon > 0$ so that, for any two points $x \neq y$ whose orbits remain in the neighborhood forever, there exists some $k \geq 0$ so that $f^{\circ k}(x)$ and $f^{\circ k}(y)$ have distance greater than ϵ. Using Sullivan's results from §16, show that this condition is satisfied if and only if f is hyperbolic. (However, a map with a parabolic fixed point may be expansive on the Julia set itself.)

Problem 19-f. Locally connected Julia sets. If a Julia set J is locally connected, show that it must be connected. Prove the following theorem of Torhorst: *If* $X \subset S^2$ *is compact and locally connected, then*

---

*   In higher dimensions, (1) is replaced by the assumption that the tangent vector bundle of the underlying manifold, restricted to Ω, splits as the direct sum of a bundle on which the derivative Df is expanding and a bundle on which it is contracting.

the boundary of every complementary component must be locally connected.
(See [Whyburn]; and compare the proof of 17.14.) Prove also that the
diameters of the complementary components tend to zero. (Here is one
possible outline: Let U be a component of $S^2 \smallsetminus X$ of diameter greater
than 4ϵ, with ϵ small. After rotating the sphere, we may assume that U
contains points at distance 2ϵ from the equator in both the northern and
southern hemispheres. Then each parallel at distance less than ϵ from the
equator must intersect U in at least one interval whose endpoints cannot be
joined within X by any connected set of diameter less than ϵ. Let δ_U be
the minimum length of such an interval. Then U has area $A(U) \geq 2\epsilon\delta_U$.
Now if there were infinitely many such components U, then their areas
would tend to zero, hence the numbers δ_U would tend to zero, and X
would not be locally connected.)

Appendix A. Theorems from Classical Analysis

This appendix will describe some miscellaneous theorems from classical complex variable theory. We first complete the arguments from 11.14, 17.3 and 18.2 by proving Jensen's inequality and the Riesz brothers' theorem. We then describe results from the theory of univalent functions, due to Gronwall and Bieberbach, in order to prove the Koebe Quarter Theorem for use in Appendix G. (By definition, a function of one complex variable is called *univalent* if it is holomorphic and injective.)

We begin with a discussion of Jensen's inequality. (J. L. W. V. Jensen was the president of the Danish Telephone Company, and a noted amateur mathematician.) Let $f : \mathbb{D} \to \mathbb{C}$ be a holomorphic function on the open disk which is not identically zero. Given any radius $0 < r < 1$, we can form the average

$$A(f,r) \;=\; \frac{1}{2\pi} \int_0^{2\pi} \log|f(re^{i\theta})|\,d\theta$$

of the quantity $\log|f(z)|$ over the circle $|z| = r$.

Lemma A.1 (Jensen's Inequality). *This average $A(f,r)$ is monotone increasing (and also convex upwards) as a function of r. Hence $A(f,r)$ either converges to a finite limit or diverges to $+\infty$ as $r \nearrow 1$.*

In fact the proof will show something much more precise.

Lemma A.2. *If we consider $A(f,r)$ as a function of $\log r$, then it is piecewise linear, with slope $dA(f,r)/d\log r$ equal to the number of roots of f inside the disk $\mathbb{D}_r$ of radius r, where each root is to be counted with its appropriate multiplicity.*

In particular, the function $A(f,r)$ is determined, up to an additive constant, by the location of the roots of f. To prove this Lemma, note first that we can write $d\theta = dz/iz$ around any loop $|z| = r$. Consider an annulus $\mathcal{A} = \{z \;;\; r_0 < |z| < r_1\}$ which contains no zeros of f. According to the "Argument Principle", the integral

$$n \;=\; \frac{1}{2\pi i} \oint_{|z|=r} d\log f(z) \;=\; \frac{1}{2\pi i} \oint \frac{f'(z)dz}{f(z)}$$

measures the number of zeros of f inside the disk $\mathbb{D}_r$. It follows that the difference $\log f(z) - \log z^n$ can be defined as a single valued holomorphic function throughout this annulus $\mathcal{A}$. Therefore, the integral of $(\log f(z) - \log z^n)dz/iz$ around a loop $|z| = r$ must be independent of r, as long as $r_0 < r < r_1$. Taking the real part, it follows that the difference $A(f,r) -$

$A(z^n, r)$ is a constant, independent of r. Since $A(z^n, r) = n \log r$, this proves that the function $\log r \mapsto A(f, r)$ is linear with slope n for $r_0 < r < r_1$.

Finally, note that the average $A(f, r)$ takes a well defined finite value even when f has one or more zeros on the circle $|z| = r$, since the singularity of $\log |f(z)|$ at a zero of f is relatively mild. (For example, the indefinite integral $\int \log |x| dx = x \log |x| - x$ is continuous as a function of x.) Continuity of $A(f, r)$ as r varies through such a singularity is not difficult, and will be left to the reader. □

Theorem A.3 (F. and M. Riesz). *Suppose that $f : \mathbb{D} \to \mathbb{C}$ is bounded and holomorphic on the open unit disk. If the radial limit*

$$\lim_{r \nearrow 1} f(re^{i\theta})$$

exists and takes some constant value c_0 for θ belonging to a set $E \subset [0, 2\pi]$ of positive Lebesgue measure, then f must be identically equal to c_0.

(Compare the discussion of Fatou's Theorem in 17.3.)

Proof. Without loss of generality, we may assume that $c_0 = 0$. Let $E(\epsilon, \delta)$ be the measurable set consisting of all $\theta \in E$ such that

$$|f(re^{i\theta})| < \epsilon \qquad \text{whenever} \qquad 1 - \delta < r < 1.$$

Evidently, for each fixed ϵ, the union of the nested family of sets $E(\epsilon, \delta)$ contains E. Therefore, the Lebesgue measure $\ell(E(\epsilon, \delta))$ must tend to a limit which is $\geq \ell(E)$ as $\delta \searrow 0$. In particular, given ϵ we can choose δ so that $\ell(E(\epsilon, \delta)) > \ell(E)/2$. Now consider the average $A(f, r)$ of Jensen's Inequality, where $r > 1 - \delta$. Multiplying f by a constant if necessary, we may assume that $|f(z)| < 1$ for all $z \in \mathbb{D}$. Thus the expression $\log |f(re^{i\theta})|$ is less than or equal to zero everywhere, and less than or equal to $\log \epsilon$ throughout a set of measure at least $\ell(E)/2$. This proves that

$$2\pi A(f, r) < \log(\epsilon) \, \ell(E)/2$$

whenever r is sufficiently close to 1. Since ϵ can be arbitrarily small, this implies that $\lim_{r \nearrow 1} A(f, r) = -\infty$, which contradicts A.1 unless f is identically zero. □

Now consider the following situation. Let K be a compact connected subset of $\mathbb{C}$, and suppose that the complement $\mathbb{C} \smallsetminus K$ is conformally diffeomorphic to the complement $\mathbb{C} \smallsetminus \overline{\mathbb{D}}$.

Lemma A.4 (Gronwall Area Formula). *Let* $\psi : \mathbb{C} \setminus \overline{\mathbb{D}} \rightarrow \mathbb{C} \setminus K$ *be a conformal isomorphism, with Laurent series expansion*

$$\psi(w) = b_1 w + b_0 + b_{-1}/w + b_{-2}/w^2 + \cdots .$$

Then the 2-dimensional Lebesgue measure of K *is given by*

$$\operatorname{area}(K) = \pi \sum_{n \leq 1} n|b_n|^2 = \pi\left(|b_1|^2 - |b_{-1}|^2 - 2|b_{-2}|^2 - \cdots\right) .$$

Proof. For any $r > 1$ consider the image under ψ of the circle $|w| = r$. This will be some embedded circle in $\mathbb{C}$ which encloses a region of area say $A(r)$. We can compute this area by Green's Theorem, as follows. Let $\psi(re^{i\theta}) = z = x + iy$. Then

$$A(r) = \oint x\, dy = -\oint y\, dx = \tfrac{1}{2i} \oint \bar{z}\, dz ,$$

to be integrated around the image of $|w| = r$. Substituting the Laurent series
$z = \sum_{n \leq 1} b_n w^n$, with $w = r^n e^{ni\theta}$, this yields

$$A(r) = \tfrac{1}{2} \sum_{m,n \leq 1} n \bar{b}_m b_n r^{m+n} \oint e^{(n-m)i\theta} d\theta .$$

Since the integral equals 2π if $m = n$, and is zero otherwise, we obtain

$$A(r) = \pi \sum_{n \leq 1} n|b_n|^2 r^{2n} .$$

Therefore, taking the limit as $r \searrow 1$, we obtain the required formula. $\quad\square$

Remark. Unfortunately, it is difficult to make any estimate on the rate of convergence of this series. Intuitively, if the set K has a complicated shape, then it seems likely that very high order terms will play an important role.

One trivial consequence of A.4 is the inequality

$$|b_1|^2 \geq \sum_1^\infty m|b_{-m}|^2 .$$

In particular we have the following.

Corollary A.5 (Gronwall Inequality). *With* K *and* $\psi(w) = \sum_{n \leq 1} b_n w^n$ *as above, we have* $|b_1| \geq |b_{-1}|$, *with equality if and only if* K *is a straight line segment.*

Proof. Since $\operatorname{area}(K) \geq 0$, we have $|b_1| \geq |b_{-1}|$. Furthermore, if equality holds then all of the remaining coefficients must be zero: $b_{-2} =$

$b_{-3} = \cdots = 0$. After a rotation of the w coordinate and a linear transformation of the $z = \psi(w)$ coordinate, the Laurent series will reduce to the simple formula $z = w + w^{-1}$. As noted in §7, this transformation carries $\mathbb{C} \setminus \overline{\mathbb{D}}$ diffeomorphically onto the complement of the interval $[-2, 2]$. □

Now consider an open set $U \subset \mathbb{C}$ which contains the origin and is conformally isomorphic to the open disk.

Lemma A.6 (Bieberbach Inequality). *If $\psi : \mathbb{D} \to U$ is a conformal isomorphism with power series expansion $\psi(\eta) = \sum_{n \geq 1} a_n \eta^n$, then $|a_2| \leq 2|a_1|$, with equality if and only if $\mathbb{C} \setminus U$ is a closed half-line pointing towards the origin.*

Remark. The Bieberbach conjecture, proved by DeBrange, asserts that $|a_n| \leq n|a_1|$ for all n. Again, equality holds if $\mathbb{C} \setminus U$ is a closed half-line pointing towards the origin, for example for $\psi(\eta) = \eta + 2\eta^2 + 3\eta^3 + \cdots = \eta/(1 - \eta)^2$.

Proof of A.6. After composing ψ with a linear transformation, we may assume that $a_1 = 1$. Let us set $\eta = 1/w^2$, so that each point $\eta \neq 0$ in $\mathbb{D}$ corresponds to two points $\pm w \in \mathbb{C} \setminus \overline{\mathbb{D}}$. Similarly, set $\psi(\eta) = \zeta = 1/z^2$, so that each $\zeta \neq 0$ in U corresponds to two points $\pm z$ in some centrally symmetric neighborhood N of infinity. A brief computation shows that ψ corresponds to a Laurent series

$$w \mapsto z = 1/\sqrt{\psi(1/w^2)} = w - \tfrac{1}{2}a_2/w + (\text{higher terms})$$

which maps $\mathbb{C} \setminus \overline{\mathbb{D}}$ diffeomorphically onto N. Thus $|a_2| \leq 2$ by Gronwall's Inequality, with equality if and only if N is the complement of a line segment, necessarily centered at the origin. Expressing this condition on N in terms of the coordinate $\zeta = 1/z^2$, we see that equality holds if and only if U is the complement of a half-line pointing towards the origin. □

Theorem A.7 (Koebe-Bieberbach). *Again suppose that the map*

$$\eta \mapsto \psi(\eta) = a_1\eta + a_2\eta^2 + \cdots$$

carries the unit disk $\mathbb{D}$ diffeomorphically onto an open set $U \subset \mathbb{C}$. Then the distance r between the origin and the boundary of U satisfies

$$\tfrac{1}{4}|a_1| \leq r \leq |a_1|.$$

Here the first equality holds if and only if $\mathbb{C} \setminus U$ is a half-line pointing towards the origin, and the second equality holds if and only if U is a disk centered at the origin.

In particular, in the special case $a_1 = 1$ the open set $U = \psi(\mathbb{D})$ necessarily contains the disk $\mathbb{D}_{1/4}$ of radius $1/4$ centered at the origin, so that $\psi^{-1} : \mathbb{D}_{1/4} \to \mathbb{D}$ is well defined and univalent. The left hand inequality was conjectured and partially proved by Koebe, and later completely proved by Bieberbach. The right hand inequality is an easy consequence of the Schwarz Lemma.

Proof of A.7. Without loss of generality, we may assume that $a_1 = 1$. If $z_0 \in \partial U$ be a boundary point with minimal distance r from the origin, then we must prove that $\frac{1}{4} \leq r \leq 1$. We will compose ψ with the linear fractional transformation $z \mapsto z/(1 - z/z_0)$ which maps z_0 to infinity. Then the composition has the form

$$\eta \mapsto \psi(\eta)/(1 - \psi(\eta)/z_0) = \eta + (a_2 + 1/z_0)\eta^2 + \cdots.$$

By Bieberbach's Theorem we have $|a_2| \leq 2$ and $|a_2 + 1/z_0| \leq 2$, hence $|1/z_0| = 1/r \leq 4$, or $r \geq 1/4$. Here equality holds only if $|a_2| = 2$ and $|a_2 + 1/z_0| = 2$. The exact description of U then follows easily.

On the other hand, suppose that $r \geq 1$. Then the inverse mapping ψ^{-1} is defined and holomorphic throughout the unit disk $\mathbb{D}$, and takes values in $\mathbb{D}$. Since its derivative at zero is 1, it follows from the Schwarz Inequality that ψ is the identity map, with $r = 1$. $\square$

Here is an interesting restatement of the Quarter Theorem. Let $ds = \rho(z)|dz|$ be the Poincaré metric on the open set U, and let $r = r(z)$ be the distance from z to the boundary of U.

Corollary A.8. *If $U \subset \mathbb{C}$ is simply connected, then the Poincaré metric $ds = \rho(z)|dz|$ on U agrees with the metric $|dz|/r(z)$ up to a factor of two in either direction. That is*

$$\frac{1}{2r(z)} \leq \rho(z) \leq \frac{2}{r(z)}$$

for all $z \in U$. Again, the left equality holds if and only if $\mathbb{C} \smallsetminus U$ is a half-line pointing towards the point $z \in U$, and the right equality holds if and only if U is a round disk centered at z.

This follows, since we can choose $\psi : \mathbb{D} \to U$ mapping the origin to any given point of U, and since the Poicaré metric at the center of $\mathbb{D}$ is $2|d\eta|$. $\square$

As an example, if U is a half-plane, then the Poincaré metric precisely agrees with the $1/r$ metric $|dz|/r$.

We conclude with a problem for the reader.

Problem A-1. Area of the Filled Julia Set. For c in the Mandelbrot set, it follows from A.4 that the area $A(c)$ of the filled Julia set for $z \mapsto z^2 + c$ can be expressed as the sum of a series $\pi \left(|b_1|^2 - |b_{-1}|^2 - 3|b_{-3}|^2 - \cdots \right)$. Using the identity

$$\psi(w^2) = (\psi(w))^2 + c$$

of (9 : 5), show that each b_n is a polynomial function of c, with $b_n = 0$ for n even. Show that the function $c \mapsto A(c)$ is upper semicontinuous. (Compare 11.15.)

Appendix B. Length-Area-Modulus Inequalities

We will first study the conformal geometry of a rectangle. Let $\overline{R} = [0, \Delta x] \times [0, \Delta y]$ be a closed rectangle with interior $R = (0, \Delta x) \times (0, \Delta y)$ in the plane of complex numbers $z = x + iy$. By a *conformal metric* on R we mean a metric of the form

$$ds = \rho(z)|dz|$$

where $z \mapsto \rho(z) > 0$ is any strictly positive real valued function which is defined and continuous throughout the open rectangle. In terms of such a metric, the *length* of a smooth curve $\gamma : (a, b) \to R$ is defined to be the integral

$$\mathbf{L}_\rho(\gamma) = \int_a^b \rho(\gamma(t))|d\gamma(t)| \,,$$

and the *area* of a region $U \subset R$ is defined to be

$$\operatorname{area}_\rho(U) = \iint_U \rho(x + iy)^2 dx\, dy \,.$$

In the special case of the Euclidean metric $ds = |dz|$, with $\rho(z)$ identically equal to 1, the subscript ρ will be omitted.

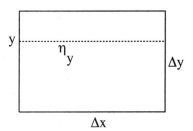

Main Lemma B.1. *If* $\operatorname{area}_\rho(R)$ *is finite, then for Lebesgue almost every* $y \in (0, \Delta y)$ *the length* $\mathbf{L}_\rho(\eta_y)$ *of the horizontal line* $\eta_y : t \mapsto (t, y)$ *at height* y *is finite. Furthermore, there exists* y *so that*

$$\frac{\mathbf{L}_\rho(\eta_y)^2}{(\Delta x)^2} \leq \frac{\operatorname{area}_\rho(R)}{\Delta x \Delta y} \,. \tag{B : 1}$$

In fact, the set consisting of all $y \in (0, 1)$ *for which this in-equality is satisfied has positive Lebesgue measure.*

Remark. Note that Δx is equal to $L(\eta_y)$, the Euclidean length, and that the product $\Delta x \Delta y$ is equal to $\operatorname{area}(R)$, the Euclidean area. Evidently the inequality (B : 1) is best possible, for in the special case of the Euclidean metric with $\rho \equiv 1$ both sides of the inequality are equal to $+1$.

Proof of B.1. We use the Schwarz inequality

$$\left(\int_a^b f(x)g(x)\, dx \right)^2 \le \left(\int_a^b f(x)^2\, dx \right) \cdot \left(\int_a^b g(x)^2\, dx \right),$$

which says (after taking a square root) that the inner product of any two vectors in the Euclidean vector space of square integrable real functions on an interval is less than or equal to the product of their norms. Taking $f(x) \equiv 1$ and $g(x) = \rho(x, y)$ for some fixed y, we obtain

$$\left(\int_0^{\Delta x} \rho(x, y)\, dx \right)^2 \le \Delta x \int_0^{\Delta x} \rho(x, y)^2\, dx,$$

or in other words

$$L_\rho(\eta_y)^2 \le \Delta x \int_0^{\Delta x} \rho(x, y)^2 dx,$$

for each constant height y. Integrating this inequality over the interval $0 < y < \Delta y$ and then dividing by Δy, we get

$$\frac{1}{\Delta y} \int_0^{\Delta y} L_\rho(\eta_y)^2 dy \le \frac{\Delta x}{\Delta y} \operatorname{area}_\rho(A). \qquad (B:2)$$

In other words, the *average* over all y in the interval $(0, \Delta y)$ of $L_\rho(\eta_y)^2$ is less than or equal to $\frac{\Delta x}{\Delta y} \operatorname{area}_\rho(A)$. Further details of the proof are straightforward. $\square$

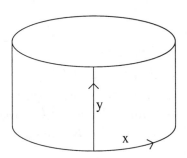

Now let us form a cylinder C of circumference Δx and height Δy by gluing the left and right edges of our rectangle together. (Alternatively, C can be described as the Riemann surface which is obtained from the infinitely wide strip $0 < y < \Delta y$ in the z-plane by identifying each point $z = x + iy$ with its translate $x + \Delta x + iy$. Compare Problem 2-f.)

Definitions. The *modulus* of such a cylinder $C = (\mathbb{R}/(\mathbb{Z}\Delta x)) \times (0, \Delta y)$

is defined to be the ratio

$$\text{mod}(C) \;=\; \Delta y / \Delta x \;>\; 0$$

of height to circumference. By the *winding number* of a closed curve γ in C we mean the integer

$$w \;=\; \frac{1}{\Delta x} \oint_\gamma dx \;.$$

Theorem B.2 (Length-Area Inequality for Cylinders).
For any conformal metric $\rho(z)|dz|$ on the cylinder C there exists some simple closed curve γ with winding number $+1$ whose length $\mathbf{L}_\rho(\gamma) = \oint_\gamma \rho(z)|dz|$ satisfies the inequality

$$\mathbf{L}_\rho(\gamma)^2 \;\le\; \text{area}_\rho(C)/\text{mod}(C) \;. \tag{B : 3}$$

Furthermore, this result is best possible: If we use the Euclidean metric $|dz|$ then

$$\mathbf{L}(\gamma)^2 \;\ge\; \text{area}(C)/\text{mod}(C) \tag{B : 4}$$

*for **every** such curve γ .*

Proof. Just as in the proof of B.1, we find a horizontal curve η_y with

$$\mathbf{L}_\rho(\eta_y)^2 \;\le\; \frac{\Delta x}{\Delta y}\text{area}_\rho(C) \;=\; \frac{\text{area}_\rho(C)}{\text{mod}(C)} \;.$$

On the other hand in the Euclidean case, for any closed curve γ of winding number one we have

$$\mathbf{L}(\gamma) \;=\; \oint_\gamma |dz| \;\ge\; \oint_\gamma dx \;=\; \Delta x \;,$$

hence $\mathbf{L}(\gamma)^2 \ge (\Delta x)^2 \;=\; \text{area}(C)/\text{mod}(C) \;.$ $\square$

Definitions. A Riemann surface A is said to be an *annulus* if it is conformally isomorphic to some cylinder. (Compare §2.) An embedded annulus $A \subset C$ is said to be *essentially embedded* if it contains a curve which has winding number one around C .

Here is an important consequence of Theorem B.2.

Corollary B.3 (An Area-Modulus Inequality). *Let $A \subset C$ be an essentially embedded annulus in the cylinder C , and suppose that A is conformally isomorphic to a cylinder C_A . Then*

$$\frac{\text{mod}(C_A)}{\text{mod}(C)} \;\le\; \frac{\text{area}(A)}{\text{area}(C)} \;\le\; 1 \;. \tag{B : 5}$$

In particular, it follows that

$$\operatorname{mod}(C_A) \leq \operatorname{mod}(C) . \tag{B : 6}$$

Proof. Let $\zeta \mapsto z$ be the embedding of C_A onto $A \subset C$. The Euclidean metric $|dz|$ on C, restricted to A, pulls back to some conformal metric $\rho(\zeta)|d\zeta|$ on C_A, where $\rho(\zeta) = |dz/d\zeta|$. According to B.2, there exists a curve γ' with winding number 1 about C_A whose length satisfies

$$L_\rho(\gamma')^2 \leq \operatorname{area}_\rho(C_A)/\operatorname{mod}(C_A) .$$

This length coincides with the Euclidean length $\mathbf{L}(\gamma)$ of the corresponding curve γ in $A \subset C$, and $\operatorname{area}_\rho(C_A)$ is equal to the Euclidean area $\operatorname{area}(A)$, so we can write this inequality as

$$L(\gamma)^2 \leq \operatorname{area}(A)/\operatorname{mod}(C_A) .$$

But according to (B : 4) we have

$$\operatorname{area}(C)/\operatorname{mod}(C) \leq L(\gamma)^2 .$$

Combining these two inequalities, we obtain

$$\operatorname{area}(C)/\operatorname{mod}(C) \leq \operatorname{area}(A)/\operatorname{mod}(C_A) ,$$

which is equivalent to the required inequality (B : 5). $\square$

Corollary B.4. *If two cylinders are conformally isomorphic, then their moduli are equal.*

Proof. If C' is conformally isomorphic to C then (B : 6) asserts that $\operatorname{mod}(C') \leq \operatorname{mod}(C)$, and similarly $\operatorname{mod}(C) \leq \operatorname{mod}(C')$. $\square$

It follows that the *modulus* of an annulus A can be defined as the modulus of any conformally isomorphic cylinder. If A is essentially embedded in some other annulus A', it then follows from (B : 6) that $\operatorname{mod}(A) \leq \operatorname{mod}(A')$.

B.5. Examples. If $\mathbb{A}_r$ is the annulus consisting of all $w \in \mathbb{C}$ with $1 < |w| < r$, then setting $z = i \log(w) \pmod{2\pi}$ we map $\mathbb{A}_r$ conformally onto a cylinder of height $\log(r)$ and circumference 2π. Hence

$$\operatorname{mod}(\mathbb{A}_r) = \log(r)/2\pi .$$

On the other hand, if we construct an annulus A from the upper half-plane $\mathbb{H}$ by identifying w with kw for some $k > 1$, then setting $z = \log(w) \pmod{\log(k)}$ we map A onto a cylinder of height π and circumference $\log(k)$. Hence

$$\operatorname{mod}(\mathbb{H}/(w \equiv kw)) = \pi/\log(k) .$$

Corollary B.6 (Grötzsch Inequality). *Suppose that $A' \subset A$ and $A'' \subset A$ are two disjoint annuli, each essentially embedded in A. Then*

$$\mathrm{mod}(A') + \mathrm{mod}(A'') \leq \mathrm{mod}(A).$$

Proof. We may assume that A is a cylinder C. According to (6) we have

$$\frac{\mathrm{mod}(A')}{\mathrm{mod}(C)} \leq \frac{\mathrm{area}(A')}{\mathrm{area}(C)}, \qquad \frac{\mathrm{mod}(A'')}{\mathrm{mod}(C)} \leq \frac{\mathrm{area}(A'')}{\mathrm{area}(C)}.$$

where all areas are Euclidean. Using the inequality

$$\mathrm{area}(A') + \mathrm{area}(A'') \leq \mathrm{area}(C),$$

the conclusion follows.   □

Now consider a *flat torus* $\mathbf{T} = \mathbb{C}/\Lambda$. Here $\Lambda \subset \mathbb{C}$ is to be a 2-dimensional *lattice*, that is an additive subgroup of the complex numbers, spanned by two elements λ_1 and λ_2 where $\lambda_1/\lambda_2 \notin \mathbb{R}$. Let $A \subset \mathbf{T}$ be an embedded annulus.

By the "*winding number*" of A in $\mathbf{T}$ we will mean the lattice element $w \in \Lambda$ which is constructed as follows. Under the universal covering map $\mathbb{C} \to \mathbf{T}$, the central curve of A lifts to a curve segment which joins some point $z_0 \in \mathbb{C}$ to a translate $z_0 + w$ by the required lattice element. We say that $A \subset \mathbf{T}$ is an *essentially embedded annulus* if $w \neq 0$.

Corollary B.7 (Bers Inequality). *If the annulus A is embedded in the flat torus $\mathbf{T} = \mathbb{C}/\Lambda$ with winding number $w \in \Lambda$, then*

$$\mathrm{mod}(A) \leq \frac{\mathrm{area}(T)}{|w|^2}. \tag{B:7}$$

Roughly speaking, if A winds many times around the torus, so that $|w|$ is large, then this annulus A must be very thin. A slightly sharper version of this inequality is given in Problem B-3 below.

Proof. Choose a cylinder C' which is conformally isomorphic to A. The Euclidean metric $|dz|$ on $A \subset \mathbf{T}$ corresponds to some metric $\rho(\zeta)|d\zeta|$ on C', with

$$\mathrm{area}_\rho(C') = \mathrm{area}(A).$$

By B.2 we can choose a curve γ' of winding number one on C', or a corresponding curve γ on $A \subset \mathbf{T}$, with

$$\mathbf{L}(\gamma)^2 = \mathbf{L}_\rho(\gamma')^2 \leq \frac{\mathrm{area}_\rho(C')}{\mathrm{mod}(C')} = \frac{\mathrm{area}(A)}{\mathrm{mod}(A)} \leq \frac{\mathrm{area}(T)}{\mathrm{mod}(A)}.$$

Now if we lift γ to the universal covering space $\mathbb{C}$ then it will join some point z_0 to z_0+w. Hence its Euclidean length $\mathbf{L}(\gamma)$ must satisfy $\mathbf{L}(\gamma) \geq |w|$. Thus

$$|w|^2 \leq \frac{\text{area}(T)}{\text{mod}(A)},$$

which is equivalent to the required inequality (8). $\square$

Now consider the following situation. Let $U \subset \mathbb{C}$ be a bounded simply connected open set, and let $K \subset U$ be a compact subset so that the difference $A = U \smallsetminus K$ is a topological annulus. As noted in §2, such an annulus must be conformally isomorphic to a finite or infinite cylinder. By definition an infinite cylinder, that is a cylinder of infinite height, has modulus zero. (Such an infinite cylinder may be either one-sided infinite, conformally isomorphic to a punctured disk, or two-sided infinite, conformally isomorphic to the punctured plane.)

Corollary B.8. *Suppose that $K \subset U$ as described above. Then K reduces to a single point if and only if the annulus $A = U \smallsetminus K$ has infinite modulus. Furthermore, the diameter of K is bounded by the inequality*

$$4\,\text{diam}(K)^2 \leq \frac{\text{area}(A)}{\text{mod}(A)} \leq \frac{\text{area}(U)}{\text{mod}(A)}. \qquad (\text{B}:8)$$

Proof. According to B.2, there exists a curve with winding number one about A whose length satisfies $L^2 \leq \text{area}(A)/\text{mod}(A)$. Since K is enclosed within this curve, it follows easily that $\text{diam}(K) \leq \mathbf{L}/2$, and the inequality (B : 8) follows. Conversely, if K is a single point then using (B : 6) we see easily that $\text{mod}(A) = \infty$. $\square$

The following ideas are due to McMullen. (Compare [Branner-Hubbard, 1992].) The *isoperimetric inequality* asserts that the area enclosed by a plane curve of length $\mathbf{L}$ is at most $\mathbf{L}^2/(4\pi)$, with equality if and only if the curve is a round circle. (See for example [Courant-Robbins].) Combining this with the argument above, we see that

$$\text{area}(K) \leq \frac{\mathbf{L}^2}{4\pi} \leq \frac{\text{area}(A)}{4\pi\,\text{mod}(A)}.$$

Writing this inequality as $4\pi\,\text{mod}(A) \leq \text{area}(A)/\text{area}(K)$ and adding $+1$ to both sides we obtain the completely equivalent inequality

$$1 + 4\pi\,\text{mod}(A) \leq \text{area}(U)/\text{area}(K),$$

or in other words

$$\text{area}(K) \ \leq \ \frac{\text{area}(U)}{1 + 4\pi \, \text{mod}(A)} \ . \tag{B : 9}$$

This can be sharpened as follows:

Corollary B.9 (McMullen Inequality). *If* $A = U \smallsetminus K$ *as above, then*

$$\text{area}(K) \ \leq \ \text{area}(U)/e^{4\pi \, \text{mod}(A)} \ .$$

Proof. Cut the annulus A up into n concentric annuli A_i, each of modulus equal to $\text{mod}(A)/n$. Let K_i be the bounded component of the complement of A_i, and assume that these annuli are nested so that $A_i \cup K_i = K_{i+1}$ with $K_1 = K$, and let $K_{n+1} = A \cup K = U$. Then $\text{area}(K_{i+1})/\text{area}(K_i) \geq 1 + 4\pi \, \text{mod}(A)/n$ by (B : 9), hence

$$\text{area}(U)/\text{area}(K) \ \geq \ (1 + 4\pi \, \text{mod}(A)/n)^n \ ,$$

where the right hand side converges to $e^{4\pi \, \text{mod}(A)}$ as $n \to \infty$. $\square$

Concluding Problems:

Problem B-1. Many Short Lines. In the situation of Lemma B.1 on the unit square $[0, 1] \times [0, 1]$, show that more than half of the horizontal curves η_y have length $\mathbf{L}_\rho(\eta_y) \leq \sqrt{2 \, \text{area}_\rho(I^2)}$. (Here "more than half" is to be interpreted in the sense of Lebesgue measure.)

Problem B-2. Defining the Modulus by Potential Theory. Recall that a real valued function u on a Riemann surface is *harmonic* if and only if it can be described locally as the real part of a complex analytic function $u + iv$. The *conjugate* real valued function v is then well defined locally up to an additive constant. With the cylinder C as in B.2, show that there is one and only one harmonic function $u : C \to \mathbb{R}$ such that $u(x + iy) \to 0$ as $x + iy$ tends to the bottom boundary $y = 0$, and $u(x + iy) \to 1$ as $x + iy$ tends to the top boundary. In fact it is given by $u(x + iy) = y/\Delta y$. If γ is a curve with winding number $+1$ around C, show that

$$i \oint_\gamma d(u + iv) \ = \ \oint_\gamma d(x + iy)/\Delta y \ = \ \frac{1}{\text{mod}(C)} \ .$$

Problem B-3. Sharper Bers Inequality. If the flat torus $\mathbb{T} = \mathbb{C}/\Lambda$ contains several disjoint annuli A_i, all with the same "winding number"

$w \in \Lambda$, show that

$$\sum \mathrm{mod}(A_i) \leq \mathrm{area}(\mathbb{T})/|w|^2 .$$

If two essentially embedded annuli are disjoint, show that they necessarily have the same winding number.

Problem B-4. Branner-Hubbard Criterion. Let $K_1 \supset K_2 \supset K_3 \supset \cdots$ be compact subsets of $\mathbb{C}$ with each K_{n+1} contained in the interior of K_n . Suppose further that each interior K_n^o is simply connected, and that each difference $A_n = K_n^o \smallsetminus K_{n+1}$ is an annulus. If $\sum_1^\infty \mathrm{mod}(A_n)$ is infinite, show that the intersection $\bigcap K_n$ reduces to a single point. Show that the converse statement is false: this intersection may reduce to a single point even though $\sum_1^\infty \mathrm{mod}(A_n) < \infty$. (As a first step, you could consider the open unit disk $\mathbb{D}$ and a closed disk $\overline{\mathbb{D}}'$ of radius $1/2$ with center $1/2 - \epsilon$, showing that $\mathrm{mod}(\mathbb{D} \smallsetminus \overline{\mathbb{D}}')$ tends to zero as $\epsilon \searrow 0$.)

Appendix C. Rotations, Continued Fractions, and Rational Approximation

The study of *recurrence* is a central topic in many parts of dynamics: *How often and how closely does an orbit return to a neighborhood of its initial point?* In the case of irrational rotations of a circle we can give a rather precise description of the answer. This description, which has its roots in classical number theory, turns out to be important not only in holomorphic dynamics, but also in celestial mechanics and other areas where "small divisor" problems occur.

Let $S^1 \subset \mathbb{C}$ be the unit circle, consisting of all complex numbers of absolute value 1. Given some fixed $\lambda \in S^1$, we are interested in the dynamical system $z \mapsto \lambda z$ for $z \in S^1$. Since any two orbits are isometric under a rotation of the circle, it suffices to study the orbit

$$1 \mapsto \lambda \mapsto \lambda^2 \mapsto \lambda^3 \mapsto \cdots .$$

We are particularly interested in the case where λ is not a root of unity, so that there is no periodic point. Here is a picture of a typical example, showing the first few points on this orbit. (To simplify the picture, each λ^k is labeled simply by k.)

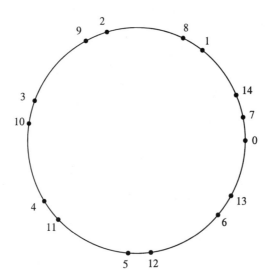

Figure 40. Successive orbit points for a rotation through an angle of $\xi \approx .148\cdots$.

Definition. We will say that the sequence $\lambda^1, \lambda^2, \lambda^3, \ldots$ has a *close return* to $\lambda^0 = 1$ at the *time* q if λ^q is closer to 1 than any of its predecessors:

$$|\lambda^k - 1| > |\lambda^q - 1| \quad \text{for every} \quad 0 < k < q \,.$$

As an example, this figure illustrates a rotation with close returns at times $q = 1, 6, 7, (27, \ldots)$.

As usual, we can equally well use the additive model $\mathbb{R}/\mathbb{Z}$ for the circle, in place of the multiplicative model $S^1 = \{\lambda \in \mathbb{C} : |\lambda| = 1\}$. These are related by letting each $\xi \in \mathbb{R}/\mathbb{Z}$ correspond to $\lambda = e^{2\pi i \xi} \in S^1$. Thus a completely equivalent problem is to study the dynamical system

$$x \mapsto x + \xi \pmod{\mathbb{Z}} \,,$$

with typical orbit

$$0 \mapsto \xi = \xi_1 \mapsto \xi_2 \mapsto \xi_3 \mapsto \cdots \,,$$

where $\xi_k = k\xi \in \mathbb{R}/\mathbb{Z}$.

Any pair of distinct points $\xi, \eta \in \mathbb{R}/\mathbb{Z}$ cuts the circle $\mathbb{R}/\mathbb{Z}$ into two arcs of total length one. It will be convenient to define the norm

$$0 \le \|\xi\| \le 1/2$$

to be the length of the shorter arc between 0 and ξ, setting $\|\xi\| = 0$ if and only if $\xi = 0$. The norm $\|\xi - \eta\|$ can be identified with the length of the shorter arc between ξ and η, or in other words with the Riemannian *distance* between ξ and η. Note the identity

$$\|\xi_i - \xi_j\| = \|\xi_{|i-j|}\| \,,$$

and note also that

$$|e^{2\pi i \xi} - 1| = 2 \sin(\pi \|\xi\|) \,,$$

with

$$4\|\xi\| \le |e^{2\pi i \xi} - 1| \le 2\pi \|\xi\| \,. \tag{C : 1}$$

Since the function $\|\xi\| \mapsto 2 \sin(\pi \|\xi\|)$ is strictly monotone, it follows that the sequence $\lambda^1, \lambda^2, \ldots$ has a close return to 1 at time q if and only if the sequence $\xi_1, \xi_2, \ldots$ has a close return to 0 at time q, so that

$$\|\xi_k\| > \|\xi_q\| \quad \text{for every} \quad 0 < k < q \,.$$

It will be convenient to number the times of these close returns as $1 = q_1 < q_2 < \cdots$ and to denote the distance from zero at the n-th close return by $d_n = \|\xi_{q_n}\|$. If ξ is irrational, evidently we obtain an infinite sequence of close return times, with

$$1/2 > \|\xi\| = d_1 > d_2 > d_3 > \cdots \to 0 \,.$$

On the other hand, if $\xi = p/q$ then this sequence must terminate as soon as we reach an m with $q_m = q$, and we have

$$1/2 \geq \|\xi\| = d_1 > d_2 > \cdots > d_{m-1} > d_m = 0 \; .$$

In either case, any ξ_k with $q_n \leq k < q_{n+1}$ must satisfy $\|\xi_k\| \geq d_n > d_{n+1}$.

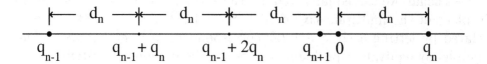

Figure 41. Locations of three successive close returns along the interval from $\xi_{q_{n-1}}$ through 0 to ξ_{q_n} within $\mathbb{R}/\mathbb{Z}$ for $n \geq 2$, illustrated for the case $a_n = 3$. Depending on the parity of n, the orientation may be reversed. As in the previous figure, each point ξ_k along the orbit is labeled simply by the integer k.

Lemma C.1. *The finite or infinite sequence of close return times determines the angle ξ up to sign. A given sequence occurs if and only if it has the form $1 = q_1 < q_2 < q_3 < \cdots$ with $q_{n+2} \equiv q_n \pmod{q_{n+1}}$ for $n \geq 1$, and with $q_{n+2} > q_n + q_{n+1}$ in case this sequence terminates with q_{n+2}.*

The proof of C.1 will be based on the classical *continued fraction algorithm*. Given any real number $0 < \hat{x} < 1$, we can construct a finite or infinite sequence a_1, a_2, a_3, ... of positive integers, as well as a finite or infinite sequence x_1, x_2, x_3, ... of remainder terms, by induction, as follows. We start with $x_0 = \hat{x}$, and set

$$1/x_{n-1} = a_{n+1} + x_{n+1} \; , \qquad (\text{C}:2)$$

with $a_{n+1} \in \mathbb{Z}$ and $0 \leq x_{n+1} < 1$. This x_{n+1} is called the *fractional part* frac$(1/x_n)$, and a_{n+1} is called the *integer part* int$(1/x_n)$. If $\hat{x}$ is irrational, then all of the x_n will be irrational, and this inductive procedure continues indefinitely. This construction is often summarized by the infinite continued fraction equation

$$\hat{x} = \cfrac{1}{a_1 + x_1} = \cfrac{1}{a_1 + \cfrac{1}{a_2 + x_2}} = \cdots = \cfrac{1}{a_1 + \cfrac{1}{a_2 + \cfrac{}{\ddots + \cfrac{1}{a_n + \ddots}}}}$$

On the other hand, if $x_0 = \hat{x}$ is a rational number p/q, then x_1 will be a rational number with denominator p strictly less than q. It follows that this procedure must terminate after at most q steps, reaching some x_n which is equal to zero, so that $x_{n-1} = 1/a_n$ while $1/x_n$ is not defined. In this case we obtain the finite continued fraction equation

$$\frac{p}{q} = \cfrac{1}{a_1 + \cfrac{1}{a_2 + \cfrac{\ddots}{\quad + \cfrac{1}{a_n}}}}$$

with $a_n \geq 2$.

Given $a_1, a_2, \ldots \geq 1$ as above, construct sequences $p_0, p_1, \ldots$ and $q_0, q_1, \ldots$ of non-negative integers by setting

$$\begin{aligned} p_0 &= 1, & p_1 &= 0, & p_{n+1} &= a_n p_n + p_{n-1}, \\ q_0 &= 0, & q_1 &= 1, & q_{n+1} &= a_n q_n + q_{n-1}. \end{aligned} \qquad (\text{C}:3)$$

Since $a_n \geq 1$, it follows that

$$p_{n+1} \geq 2p_{n-1}, \qquad q_{n+1} \geq 2q_{n-1}$$

for $n \geq 2$. Thus the sequences $\{p_n\}$ and $\{q_n\}$ tend to infinity at least exponentially fast as $n \to \infty$. Using matrix notation, we can also write the equations $(\text{C}:3)$ as

$$\begin{aligned} &\begin{bmatrix} p_n & q_n \\ p_{n+1} & q_{n+1} \end{bmatrix} \\ &= \begin{bmatrix} 0 & 1 \\ 1 & a_n \end{bmatrix} \begin{bmatrix} p_{n-1} & q_{n-1} \\ p_n & q_n \end{bmatrix} = \begin{bmatrix} 0 & 1 \\ 1 & a_n \end{bmatrix} \begin{bmatrix} 0 & 1 \\ 1 & a_{n-1} \end{bmatrix} \cdots \begin{bmatrix} 0 & 1 \\ 1 & a_1 \end{bmatrix}. \end{aligned}$$
$$(\text{C}:3')$$

It follows that the determinant of this matrix is given by

$$p_n q_{n+1} - q_n p_{n+1} = (-1)^n . \qquad (\text{C}:4)$$

In particular, p_n and q_n are relatively prime, so that each p_n/q_n is a fraction in lowest terms.

Setting $d_n = x_0 x_1 \cdots x_{n-1}$, the equation $(\text{C}:2)$ multiplied by d_n takes the form

$$d_{n-1} = a_n d_n + d_{n+1}, \qquad (\text{C}:2')$$

with $d_0 = 1$, $d_1 = x_0$. In order to get another expression for d_n, it is convenient to set $\epsilon_n = (-1)^n d_n$. Then $(\text{C}:2')$ takes the form

$$\epsilon_{n+1} = a_n \epsilon_n + \epsilon_{n-1} .$$

Comparing this equation with $(\text{C}:3)$, a straightforward induction shows that

$$\epsilon_n = p_n - q_n x_0 .$$

In particular, we have $d_n = |p_n - q_n x_0|$. It follows that we can write

$$x_0 = \frac{p_n}{q_n} - \frac{\epsilon_n}{q_n} \quad \text{with} \quad \epsilon_n = (-1)^n x_0 \cdots x_{n-1} . \qquad (C:5)$$

Since $0 < x_i < 1$, it follows that

$$0 = \frac{p_1}{q_1} < \frac{p_3}{q_3} < \frac{p_5}{q_5} < \cdots \quad < x_0 < \quad \cdots < \frac{p_6}{q_6} < \frac{p_4}{q_4} < \frac{p_2}{q_2} < 1 .$$

We can estimate the size of the error term $d_n/q_n = |\epsilon_n|/q_n$ by the computation

$$\frac{d_{n+1}}{q_{n+1}} + \frac{d_n}{q_n} = \left| \frac{p_{n+1}}{q_{n+1}} - x_0 \right| + \left| x_0 - \frac{p_n}{q_n} \right| = \left| \frac{p_{n+1}}{q_{n+1}} - \frac{p_n}{q_n} \right| = \frac{1}{q_n q_{n+1}} ,$$

using $(C:5)$ and $(C:4)$. In particular we have

$$\frac{d_n}{q_n} < \frac{d_n}{q_n} + \frac{d_{n+1}}{q_{n+1}} = \frac{1}{q_n q_{n+1}} < 2 \frac{d_n}{q_n} ,$$

or in other words

$$d_n < \frac{1}{q_{n+1}} < 2 d_n . \qquad (C:6)$$

It follows that d_n tends to zero at least exponentially fast as $n \to \infty$. In particular, it certainly follows that p_n/q_n converges to x_0.

Now let $\|q x_0\|$ be the distance from $q x_0$ to $\mathbb{Z}$, that is, the minimum over all integers p of the absolute value $|q x_0 - p|$.

> **Theorem C.2.** *For any n with $1 \le q_n < q_{n+1}$, we can characterize q_{n+1} as the smallest integer $q > 0$ such that $\|q x_0\| < \|q_n x_0\|$.*

The proof will be based on the following remark.

> **Lemma C.3.** *Let $\Lambda \subset \mathbb{R}^2$ be the additive subgroup generated by a vector $\mathbf{v} = (v_1, v_2)$ in the first quadrant and a vector $\mathbf{w} = (w_1, w_2)$ in the second quadrant, so that $w_1 < 0 < v_1$, v_2, w_2. Then no vector $\mathbf{u} = (u_1, u_2) \in \Lambda$ can lie inside the rectangle defined by the four inequalities*
>
> $$w_1 < u_1 < v_1 \quad \text{and} \quad 0 < u_2 < v_2 + w_2 . \qquad (C:7)$$

For example we can see this by setting $\mathbf{u} = a\mathbf{v} + b\mathbf{w}$, and dividing the integer (a, b)-plane into four quadrants according as $a \le 0$ or $a \ge 1$ and

$b \leq 0$ or $b \geq 1$. It is easy to check that each of these four inequalities rules out one of the four quadrants, so that there is no simultaneous solution. $\square$

Proof of C.2. Taking n even to fix our ideas, let $\mathbf{v} = \mathbf{v}_n$ and $\mathbf{w} = \mathbf{v}_{n+1} - \mathbf{v}_n$ where

$$\mathbf{v}_n = (\epsilon_n, q_n) = (p_n - q_n x_0, q_n).$$

Then $\mathbf{v}$ clearly belongs to the first quadrant, while $\mathbf{w} = (\epsilon_{n+1} - \epsilon_n, q_{n+1} - q_n)$ belongs to the second quadrant. For these vectors, the inequalities (C : 7) take the form

$$\epsilon_{n+1} - \epsilon_n < u_1 < \epsilon_n \quad \text{and} \quad 0 < u_2 < q_{n+1}.$$

On the other hand, using (C : 4) we see that the lattice Λ generated by $\mathbf{v}$ and $\mathbf{w}$ consists of all pairs $(p - qx_0, q)$ with $p, q \in \mathbb{Z}$. This proves that there is no $(p, q) \in \mathbb{Z}^2$ with

$$\epsilon_{n+1} - \epsilon_n < p - qx_0 < \epsilon_n \quad \text{and} \quad 0 < q < q_{n+1}.$$

Since $|\epsilon_{n+1} - \epsilon_n| > |\epsilon_n|$, it follows that any (p, q) with $0 < q < q_{n+1}$ must satisfy

$$|p - qx_0| \geq |\epsilon_n|,$$

as required. The proof for n odd is the same, but with the roles of $\mathbf{v}$ and $\mathbf{w}$ interchanged. $\square$

Proof of C.1. Given $\xi \in \mathbb{R}/\mathbb{Z}$, let $x_0 = \|\xi\| \in [0, 1/2]$. Then the continued fraction construction yields integers $a_n \geq 1$, and fractions p_n/q_n converging to $\|\xi\|$, where $d_n = \|q_n \xi\|$ has the order of magnitude of $1/q_{n+1}$ by (C : 6). Note that $a_1 = \text{int}(1/x_0) \geq 2$, since $x_0 \leq 1/2$, and also that $a_n \geq 2$ if the sequence of a_i terminates with a_n The integers $1 = q_1 < a_1 = q_2 < q_3 < \cdots$ are the close return times by C.2. Conversely, given the q_n we can solve for the a_n and p_n, and then verify that $\xi = \lim_{n \to \infty} p_n/q_n$ has continued fraction expansion with these coefficients, and hence has the required sequence of close return times. $\square$

Remark. This proof that the q_n are precisely the close return times depends on our choice of $x_0 = \|\xi\| \leq 1/2$. The number $x_0' = 1 - x_0$ would have the same close return times, but a very slightly different continued fraction expansion. In fact if $0 < x_0 < 1/2$ then the integers q_n and q_n' for x_0 and x_0' are related by the equation $q_{n+1}' = q_n$ for $n \geq 1$, with $q_1' = 1$.

The subject of "best" rational approximations to an irrational number $x = x_0$ is closely related. For example if $0 < q < q_n$ then dividing the inequality $|qx - p| > |q_n x - p_n|$ by $q < q_n$ we obtain

$$\left| x - \frac{p}{q} \right| > \left| x - \frac{p_n}{q_n} \right| .$$

In other words, p_n/q_n is the best approximation to x among fractions with denominator at most q_n.

Recall from §11 that an irrational number x is *Diophantine of order k* if there exists $\epsilon > 0$ so that

$$\left| x - \frac{p}{q} \right| > \frac{\epsilon}{q^k}$$

for all rational numbers p/q, or in other words if the infimum of the product

$$q^{k-1} |qx - p| ,$$

as p and q range over integers with $q > 0$, is strictly positive. Since

$$q^{k-1} |qx - p| > q_n^{k-1} |q_n x - p_n|$$

whenever $q_n < q < q_{n+1}$ by C.2, this is equivalent to the requirement that the products $q_n^{k-1} |q_n x - p_n|$ are bounded away from zero, or, using (C : 6), that the ratios q_{n+1}/q_n^{k-1} are bounded from above. Let $\mathcal{D}(k) \subset \mathbb{R} \setminus \mathbb{Q}$ be the set of all irrational numbers which are Diophantine of order k, and let

$$\mathcal{D}(2+) = \bigcap_{k>2} \mathcal{D}(k) , \qquad \mathcal{D}(\infty) = \bigcup_{k<\infty} \mathcal{D}(k) .$$

Thus $\mathcal{D}(k) \subset \mathcal{D}(\ell)$ whenever $k < \ell$, and

$$\mathcal{D}(2) \subset \mathcal{D}(2+) \subset \mathcal{D}(\infty) .$$

Lemma C.4. *This complement $(\mathbb{R}/\mathbb{Z}) \setminus \mathcal{D}(k)$ has Hausdorff dimension $\leq 2/k$. Hence the set $(\mathbb{R}/\mathbb{Z}) \setminus \mathcal{D}(\infty)$ of Liouville numbers has Hausdorff dimension zero.*

Proof. As in the proof of 11.7, if $\xi \notin \mathcal{D}(k)$, then for every $\epsilon > 0$ there exists a fraction p/q with $|\xi - p/q| \leq \epsilon/q^k$. That is, ξ belongs to a union of intervals of length $2\epsilon/q^k$. For each fixed q there are q different choices for $p/q \pmod{\mathbb{Z}}$, so the union of the corresponding intervals has total length $2\epsilon/q^{k-1}$. Summing over q, the total length is at most $2\epsilon \sum 1/q^{k-1}$, which is finite, and tends to zero as $\epsilon \to 0$, provided that $k > 2$.

Now, instead of adding the lengths of these intervals, suppose that we add the lengths raised to some power d. If $d > 2/k$, then a similar argument shows that this sum $\sum q(2\epsilon/q^k)^d = (2\epsilon)^d \sum q^{1-kd}$ is finite, and tends to zero as $\epsilon \to 0$. Therefore, $\mathbb{R}/\mathbb{Z} \setminus \mathcal{D}(k)$ has Hausdorff dimension

$\leq 2/k$. Taking the intersection as $k \to \infty$, it follows that $\mathbb{R}/\mathbb{Z} \smallsetminus \mathcal{D}(\infty)$ has Hausdorff dimension zero. $\square$

Here is a complementary statement.

Lemma C.5. *The set $\mathcal{D}(2)$ of numbers of bounded type has measure zero.*

(For a much sharper statement, see Problem C-4.) The proof of C.5 will be based on some elementary number theory. Let us say that two rational numbers $p/q < p'/q'$ are *Farey neighbors* if no rational number r/s which lies strictly between p/q and p'/q' has denominator $s \leq \max(q, q')$. Equivalently, they are Farey neighbors if the determinant $\Delta = p'q - q'p$ is equal to $+1$, so that

$$\frac{p'}{q'} - \frac{p}{q} = \frac{1}{qq'} \, . \tag{C : 8}$$

[To see this, consider the lattice ($=$ discrete additive subgroup) in the plane $\mathbb{R}^2$ which is spanned by the vectors (q, p) and (q', p'), of slope p/q and p'/q' respectively. The parallelogram P consisting of all $\alpha(p, q) + \beta(p', q')$ with $0 \leq \alpha, \beta < 1$ forms a fundamental domain for this lattice; that is, its translates cover the plane without overlap. The area of this fundamental domain P is equal to the determinant Δ, and this is also equal to the number of integer points (s, r) in P. Thus P contains an interior point (s, r) with integer coordinates if and only if $\Delta > 1$. The slope r/s corresponding to such an interior point must lie strictly between p/q and p'/q'. Furthermore, replacing α, β by $1 - \alpha, 1 - \beta$ if necessary, we may assume that $s \leq (q + q')/2 \leq \max(q, q')$.]

Proof of C.5. Let I be any interval of real numbers with rational endpoints and with length $\ell(I)$. Given any rational $0 < \epsilon < 1$, we will construct a union of subintervals of total length $\leq (1 - \epsilon)\,\ell(I)$ so that any $\xi \in I$ which satisfies

$$(*_\epsilon) \qquad\qquad |\xi - p/q| > \epsilon/q^2 \qquad \text{for all} \qquad p/q$$

must belong to one of these subintervals. Repeating the same construction for each of these subintervals, and iterating n times, it will follow that the Lebesgue measure of the set of $\xi \in I$ which satisfy $(*_\epsilon)$ is at most $(1-\epsilon)^n \ell(I)$. Since this tends to zero as $n \to \infty$, it will follow that $I \cap \mathcal{D}(2)$ has measure zero.

The construction follows. First consider the case where the endpoints p/q and p'/q' of the interval I are Farey neighbors, so that $\ell(I) = 1/(qq')$. Let I' be the subinterval which is obtained from I by removing the ϵ/q^2 neighborhood of p/q and the $\epsilon/(q')^2$ neighborhood of p'/q'.

Clearly any $\xi \in I$ which satisfies (10_ϵ) must belong to I', which is either vacuous or has length

$$\ell(I') = \ell(I) - \epsilon/q^2 - \epsilon/(q')^2 \le \ell(I)(1-\epsilon) ,$$

as required.

For the more general case, where $\ell(I) > 1/qq'$, we proceed as follows. Let m be the maximum of q and q', and consider the *Farey series* consisting of all fractions p''/q'' in the interval I with denominator $q'' \le m$. These points cut the interval I up into subintervals $I_1, \ldots, I_N$, where the endpoints of each I_j are evidently Farey neighbors. Applying the construction above to each I_j, the conclusion follows. $\square$

Concluding Problems:

Problem C-1. In the simplest possible case $a_1 = a_2 = \cdots = 1$, show that

$$\{q_n\} = \{p_{n+1}\} = \{0, 1, 1, 2, 3, 5, 8, 13, 21, \ldots\} ,$$

yielding the sequence of Fibonacci numbers. Prove the asymptotic formula $q_n \sim \gamma^n/\sqrt{5}$ as $n \to \infty$, and hence $p_n/q_n \to 1/\gamma$ as $n \to \infty$, where $\gamma = (\sqrt{5}+1)/2$. Show that this special case corresponds to the slowest possible growth for the sequences $\{p_n\}$ and $\{q_n\}$.

Problem C-2. Show that $x \in \mathbb{R} \setminus \mathbb{Q}$ satisfies a quadratic polynomial equation with integer coefficients if and only if the entries $a_1, a_2, \ldots$ of its continued fraction expansion are eventually periodic.

Problem C-3. Define the Euler polynomials

$$\mathcal{P}() = 1, \quad \mathcal{P}(x) = x, \quad \mathcal{P}(x,y) = 1 + xy ,$$
$$\mathcal{P}(x,y,z) = x + z + xyz , \quad \ldots .$$

by setting $\mathcal{P}(x_1, x_2, \ldots, x_n)$ equal to the sum of all distinct monomials in the variables $x_1, \ldots, x_n$ which can be obtained from the product $x_1 x_2 \cdots x_n$ by striking out any number of consecutive pairs. Note that $\mathcal{P}(x_1, x_2, \ldots, x_n) = \mathcal{P}(x_n, x_{n-1}, \ldots, x_1)$, and show that

$$\mathcal{P}(x_1, \ldots, x_{n+1}) = \mathcal{P}(x_1, \ldots, x_{n-1}) + \mathcal{P}(x_1, \ldots, x_n) x_{n+1}$$

for $n \ge 1$. For any continued fraction, show that the numerators p_n and denominators q_n can be expressed as polynomial functions of the entries a_i by the formulas

$$p_{n+1} = \mathcal{P}(a_2, \ldots, a_n) , \qquad q_{n+1} = \mathcal{P}(a_1, \ldots, a_n) ,$$

so that

$$\cfrac{1}{a_1 + \cfrac{1}{a_2 + \cfrac{\ddots}{ + \cfrac{1}{a_{n-1} + \cfrac{1}{a_n}}}}} = \frac{P(a_2, \ldots, a_n)}{P(a_1, \ldots, a_n)} .$$

Problem C-4. Define the *Gauss map* $g : (0, 1] \to (0, 1]$ by the requirement that $g(x) \equiv 1/x \pmod{\mathbb{Z}}$. Show that the probability measure

$$\mu(S) = \frac{1}{\log 2} \int_S \frac{dx}{1+x}$$

on $(0, 1]$ is g-invariant, in the sense that $\mu \circ g^{-1} = \mu$. It turns out that this measure is ergodic, that is every measurable g-invariant subset $S = g^{-1}(S)$ must have measure either zero or one. (For a proof, see [Cornfeld, Fomin and Sinai].) Assuming this, consider the set S_k consisting of all irrational numbers $x \in (0, 1)$ such that the integer k occurs at most finitely often among the entries a_i of the continued fraction expansion for x, and show that $\mu(S_k) = 0$. In fact, using the Birkhoff Ergodic Theorem, prove the following more precise statement. For Lebesgue almost every x, the frequency

$$\lim_{n \to \infty} \frac{1}{n} \sum_{\{i \le n \; ; \; a_i = k\}} 1$$

of occurrences of some given integer k among the a_i is well defined and equal to

$$\mu \left(\frac{1}{k+1}, \frac{1}{k} \right) = \log_2 \left(\frac{1 + 1/k}{1 + 1/(k+1)} \right) > 0 .$$

Appendix D. Remarks concerning Two Complex Variables

Many of the arguments in these lectures are strictly one-dimensional. In fact several of our underlying principles break down completely in the two variable case.

In order to illustrate these differences, it is useful to consider the family of generalized *Hénon maps*, which can be described as follows. (Compare Friedland and Milnor.) Choose a complex constant $\delta \neq 0$ and a polynomial map $f : \mathbb{C} \to \mathbb{C}$ of degree $d \geq 2$, and consider doubly infinite sequences of complex numbers $\dots, z_{-1}, z_0, z_1, z_2, \dots$ satisfying the recurrence relation

$$z_{n+1} - f(z_n) + \delta\, z_{n-1} = 0.$$

Evidently we can solve for (z_n, z_{n+1}) as a polynomial function $F(z_{n-1}, z_n)$, where the transformation $F(z_{n-1}, z_n) = (z_n, f(z_n) - \delta z_{n-1})$ has Jacobian matrix

$$\begin{bmatrix} 0 & 1 \\ -\delta & f'(z_n) \end{bmatrix}$$

with constant determinant δ, and with trace $f'(z_n)$. Similarly we can solve for (z_{n-1}, z_n) as a polynomial function $F^{-1}(z_n, z_{n+1})$ with Jacobian determinant δ^{-1}. As an example, consider the quadratic polynomial $f(z) = z^2 + \lambda z$, having a fixed point of multiplier λ at the origin. Then

$$F(x, y) = (y, y^2 + \lambda y - \delta x), \tag{D:1}$$

where x and y are complex variables. This map has a fixed point at $(0,0)$, where the eigenvalues λ_1 and λ_2 of the Jacobian matrix satisfy $\lambda_1 + \lambda_2 = \lambda$ and $\lambda_1 \lambda_2 = \delta$. Evidently, by the appropriate choice of λ and δ, we can realize any desired non-zero λ_1 and λ_2. If $\lambda_1 \neq \lambda_2$, then we can diagonalize this Jacobian matrix by a linear change of coordinates.

> **Lemma D.1.** *Consider any holomorphic transformation $F(x,y) = (x', y')$ in two complex variables, with*
> $$x' = \lambda_1 x + \mathcal{O}(|x^2| + |y^2|), \quad y' = \lambda_2 y + \mathcal{O}(|x^2| + |y^2|)$$
> *as $(x, y) \to (0,0)$. If the eigenvalues λ_1 and λ_2 of the derivative at the origin satisfy $1 > |\lambda_1| \geq |\lambda_2| > |\lambda_1^2|$, then F is conjugate, under a local holomorphic change of coordinates, to the linear map $L(u, v) = (\lambda_1 u, \lambda_2 v)$.*

Proof. We must show that there exists a change of coordinates $(u, v) = \phi(x, y)$, defined and holomorphic throughout a neighborhood of the origin, so that $\phi \circ F \circ \phi^{-1} = L$. As in the proof of the Kœnigs Theorem, 8.2, we first choose a constant c so that $1 > c > |\lambda_1| \geq |\lambda_2| > c^2$. To any orbit

$$(x_0, y_0) \overset{F}{\mapsto} (x_1, y_1) \overset{F}{\mapsto} \cdots$$

near the origin, we associate the sequence of points

$$(u_n, v_n) = L^{-n}(x_n, y_n) = (x_n/\lambda_1^n, y_n/\lambda_2^n)$$

and show, using Taylor's Theorem, that it converges geometrically to the required limit $\phi(x_0, y_0)$, with successive differences bounded by a constant times $(c^2/\lambda_2)^n$. Details will be left to the reader. $\square$

Remarks. Some such restriction on the eigenvalues is essential. As an example, for the map

$$f(x, y) = (\lambda x, \lambda^2 y + x^2),$$

with eigenvalues λ and λ^2, there is no such local holomorphic change of coordinates. (Problem D-1.) For a much more precise statement as to when linearization is possible, compare Zehnder.

Now consider a Hénon map $F : \mathbb{C}^2 \overset{\approx}{\to} \mathbb{C}^2$ as in (D : 1) above, with eigenvalues satisfying the conditions of D.1. Let Ω be the attractive basin of the origin. We claim that ϕ extends to a global diffeomorphism $\Phi :$ $\Omega \overset{\approx}{\to} \mathbb{C}^2$. In fact, for any $(x, y) \in \Omega$ we set $\Phi(x, y) = L^{-n} \circ \phi \circ F^{\circ n}(x, y)$. If n is sufficiently large, then $F^n(x, y)$ is close to the origin, so that this expression is defined. Similarly, $\Phi^{-1}(u, v) = F^{-n} \circ \phi^{-1} \circ L^{\circ n}$ is well defined for large n. This shows that Φ is a holomorphic diffeomorphism with holomorphic inverse.

Note that this basin Ω is not the entire space $\mathbb{C}^2$. For example, if $|z_1|$ is sufficiently large compared with $|z_0|$, and if

$$(z_0, z_1) \overset{F}{\mapsto} (z_1, z_2) \overset{F}{\mapsto} (z_2, z_3) \overset{F}{\mapsto} \cdots$$

then it is not difficult to check that $|z_1| < |z_2| < |z_3| < \cdots$, so that (z_0, z_1) is not in Ω. *Thus we have constructed:*

(1) *a proper subset* $\Omega \subset \mathbb{C}^2$ *which is analytically diffeomorphic to all of* $\mathbb{C}^2$, *and*

(2) *a non-linear map with an attractive basin which contains no critical points.*

Evidently neither phenomenon can occur in one complex variable. Open sets satisfying (1) are called *Fatou-Bieberbach domains* since they were first constructed by Fatou, by an easy argument similar to that given here, and then later independently by Bieberbach, who had a much more difficult construction.

The proof that there are only finitely many attracting cycles also breaks down in two variables. Compare [Newhouse]. There has been a great deal of progress in the study of polynomial automorphisms of $\mathbb{C}^2$ in recent years. See for example [Bedford] and [Hubbard], as well as [Bedford, Lyubich, Smillie] and [Hubbard, Oberstvorth].

Problem D-1. For the map $F(x, y) = (\lambda x, \lambda^2 y + x^2)$, where $\lambda \neq 0, 1$, show that there is only one smooth F-invariant curve through the origin, namely $x = 0$. By way of contrast, for the associated linear map $L(x, y) = (\lambda x, \lambda^2 y)$ note that there are infinitely many F-invariant curves $y = cx^2$. Conclude that F is not locally holomorphically conjugate to a linear map.

Appendix E. Branched Coverings and Orbifolds

This will be an outline, without proofs, of definitions and results due to Thurston. (See Douady and Hubbard 1993.) We will use "branch point" as a synonym for "critical point" and "ramified point" as a synonym for "critical value". Thus if $f(z_0) = w_0$ with $f'(z_0) = 0$, then z_0 is called a *branch point* and the image $f(z_0) = w_0$ is called a *ramified point*. More precisely, if

$$f(z) = w_0 + c(z - z_0)^n + \text{(higher terms)},$$

with $n \geq 1$ and $c \neq 0$, then the integer $n = n(z_0)$ is called the *branch index* or the *local degree* of f at the point z_0. Thus $n(z) \geq 2$ if z is a branch point, and $n(z) = 1$ otherwise.

A holomorphic map $p : S' \to S$ between Riemann surfaces is called a *covering map* if each point of S has a connected neighborhood U which is *evenly covered*, in that each connected component of $p^{-1}(U) \subset S'$ maps onto U by a conformal isomorphism. A map $p : S' \to S$ is *proper* if the inverse image $p^{-1}(K)$ of any compact subset of S is a compact subset of S'. Note that every proper map is finite-to-one, and has a well defined finite degree $d \geq 1$. Such a map may also be called a *d-fold branched covering*. On the other hand, a covering map may well be infinite-to-one. Combining these two concepts, we obtain the following more general concept.

Definition. A holomorphic map $p : S' \to S$ between Riemann surfaces will be called a *branched covering map* if every point of S has a connected neighborhood U so that each connected component of $p^{-1}(U)$ maps onto U by a proper map.

Such a branched covering is said to be *regular* or *normal* if there exists a group Γ of conformal automorphisms of S', so that two points z_1 and z_2 of S' have the same image in S if and only if there is a group element γ with $\gamma(z_1) = z_2$. In this case we can identify S with the quotient manifold S'/Γ. In fact it is not difficult to check that the conformal structure of such a quotient manifold is uniquely determined. This Γ is called the group of *deck transformations* of the covering.

Regular branched covering maps have several special properties. For example, each ramified point is isolated, so that the set of all ramified points is a discrete subset of S. Furthermore, the branch index $n(z)$ depends only on the target point $f(z)$, that is, $n(z_1) = n(z_2)$ whenever $f(z_1) = f(z_2)$. Thus we can define the *ramification function* $\nu : S \to \{1, 2, 3, \ldots\}$ by setting $\nu(w)$ equal to the common value of $n(z)$ for all points z in the pre-image $f^{-1}(w)$. By definition, $\nu(w) \geq 2$ if w is a ramified point, and $\nu(w) = 1$ otherwise.

Definition. A pair (S, ν) consisting of a Riemann surface S and a "ramification function" $\nu : S \to \{1, 2, 3, \ldots\}$ which takes the value $\nu(w) = 1$ except at isolated points will be called a Riemann surface *orbifold*.

(Remark: Thurston's general concept of orbifold involves a structure which is locally modeled on the quotient of a coordinate space by a finite group. However, in the Riemann surface case only cyclic groups can occur, so a simpler definition can be used.)

Definition. If S' is simply connected, then a regular branched covering $p : S' \to S$ with ramification function ν will be called the *universal covering* for the orbifold (S, ν). We will use the notation $\tilde{S}_\nu \to (S, \nu)$ for this universal branched covering. The associated group Γ of deck transformations is called the *fundamental group* $\pi_1(S, \nu)$ of the orbifold.

> **Lemma E.1.** *With the following exceptions, every Riemann surface orbifold (S, ν) has a universal covering surface $\tilde{S}_\nu$ which is unique up to conformal isomorphism over S. The only exceptions are given by:*
>
> (1) *a surface $S \approx \hat{\mathbb{C}}$ with just one ramified point, or*
>
> (2) *a surface $S \approx \hat{\mathbb{C}}$ with two ramified points for which $\nu(w_1) \neq \nu(w_2)$.*
>
> *In these exceptional cases, no such universal covering exists.*

By definition, the *Euler characteristic* of an orbifold (S, ν) is the rational number

$$\chi(S, \nu) = \chi(S) + \sum \left(\frac{1}{\nu(w_j)} - 1\right),$$

to be summed over all ramified points, where $\chi(S)$ is the usual Euler characteristic of S. Intuitively speaking, each ramified point w_j makes a contribution of $+1$ to the usual Euler characteristic $\chi(S)$, but a smaller contribution of $1/\nu(w_j)$ to the orbifold Euler characteristic. Thus $\chi(S, \nu) < \chi(S) \leq 2$, or more precisely

$$\chi(S) - r < \chi(S, \nu) \leq \chi(S) - r/2$$

where r is the number of ramified points. *As an example, if $\chi(S, \nu) \geq 0$, with at least one ramified point, then it follows that $\chi(S) > 0$, so the base surface S can only be $\mathbb{D}$, $\mathbb{C}$ or $\hat{\mathbb{C}}$, up to isomorphism.* Compare E.5 below.

If there are infinitely many ramified points, note that we must set $\chi(S, \nu) = -\infty$. Similarly, if S is a connected surface which is not of finite type, then $\chi(S, \nu) = \chi(S) = -\infty$ by definition.

If S' and S are provided with ramification functions μ and ν respectively, then a branched covering map $f : S' \to S$ is said to yield a *covering map* $(S', \mu) \to (S, \nu)$ *between orbifolds* if the identity

$$n(z)\mu(z) = \nu(f(z))$$

is satisfied for all $z \in S'$, where $n(z)$ is the branch index. As an example, the universal covering map $\tilde{S}_\nu \to (S, \nu)$ is always a covering map of orbifolds, where $\tilde{S}_\nu$ is provided with the trivial ramification function $\mu \equiv 1$.

Lemma E.2. $f : (S', \mu) \to (S, \nu)$ *is a covering map between orbifolds if and only if it lifts to a conformal isomorphism from the universal covering $\tilde{S}'_\mu$ onto $\tilde{S}_\nu$. If f is a covering in this sense, and has finite degree d, then the Riemann-Hurwitz formula of §5.1 takes the form*

$$\chi(S', \mu) = \chi(S, \nu)d.$$

In particular, if the universal covering of (S, ν) is a covering of finite degree d, then $\chi(\tilde{S}_\nu) = \chi(S, \nu)d$.

The fundamental group and the Euler characteristic are related to each other as follows.

Lemma E.3. *Let (S, ν) be any Riemann surface orbifold which possesses a universal covering. Then:*

$\chi(S, \nu) > 0$ *if and only if the fundamental group $\pi_1(S, \nu)$ is finite.*

$\chi(S, \nu) = 0$ *if and only if the fundamental group contains either $\mathbb{Z}$ or $\mathbb{Z} \oplus \mathbb{Z}$ as a subgroup of finite index.*

$\chi(S, \nu) < 0$ *if and only if the fundamental group contains a non-abelian free product $\mathbb{Z} * \mathbb{Z}$, and hence does not contain any abelian subgroup of finite index.*

The Euler characteristic and the geometry of $\tilde{S}_\nu$ are related as follows.

Lemma E.4. *In most cases, the Euler characteristic $\chi(S, \nu)$ is positive, negative, or zero according as the universal covering $\tilde{S}_\nu$ is conformally spherical, hyperbolic, or Euclidean. The only exceptions are $\mathbb{C}$ or $\mathbb{D}$ with at most one ramified point, or an annulus or punctured disk with no ramified points.*

Remark. This statement is closely related to the Gauss-Bonnet Theorem,

$$\iint K \, dA = 2\pi \chi(S, \nu),$$

which holds for any orbifold metric (§19) which is complete with finite area, and which is sufficiently well behaved near infinity in the non-compact case. Here K is the Gaussian curvature of (S, ν) and dA is its area element.

Example. If $S = \hat{\mathbb{C}}$ with four ramified points of index $\nu(w_j) = 2$, then the torus T described in §7 provides a regular 2-fold branched covering. Its universal covering $\tilde{T} \cong \mathbb{C}$ can be identified with the universal covering of $(\hat{\mathbb{C}}, \nu)$. In this case, the Euler characteristic $\chi(\hat{\mathbb{C}}, \nu)$ is zero.

Remark E.5. There are relatively few cases in which $\chi(S, \nu) \geq 0$. In fact all of these cases can be listed quite explicitly as follows. The unramified cases are very well known, namely the sphere, plane or disk with $\chi > 0$; and the punctured plane or disk, and the infinite families consisting of annuli and tori, all with $\chi = 0$.

By the "ramification indices" we will mean the list of values of the ramification function at the r ramified points, ordered for example so that $\nu(w_1) \leq \cdots \leq \nu(w_r)$.

If $\chi(\hat{\mathbb{C}}, \nu) > 0$ with $r > 0$, then the ramification indices must be either (n, n) or $(2, 2, n)$ for some $n \geq 2$, or $(2, 3, 3)$, $(2, 3, 4)$ or $(2, 3, 5)$. These five possibilities correspond to the five types of finite rotation groups of the 2-sphere; namely to the cyclic, dihedral, tetrahedral, octahedral and icosahedral groups respectively. (Compare Milnor 1975, p. 179.)

If $\chi(\hat{\mathbb{C}}, \nu) = 0$, then the ramification indices must be either $(2, 4, 4)$, $(2, 3, 6)$, $(3, 3, 3)$ or $(2, 2, 2, 2)$. These correspond to the automorphism groups of the tilings of $\mathbb{C}$ by squares, equilateral triangles, alternately colored equilateral triangles, and parallelograms respectively. In the parallelogram case, note that there is actually a one complex parameter family of distinct possible shapes, corresponding to the cross-ratio of the four ramified points.

Similarly, if $\chi(\mathbb{C}, \nu)$ or $\chi(\mathbb{D}, \nu)$ is strictly positive, then we must have $r \leq 1$; while if $\chi(\mathbb{C}, \nu)$ or $\chi(\mathbb{D}, \nu)$ is zero, then we must have $r = 2$ with ramification indices $(2, 2)$. This is the complete list.

Concluding Problems.

Problem E-1. If $S = \mathbb{C}$ with two ramified points $\nu(1) = \nu(-1) = 2$, show that the map $z \mapsto \cos(2\pi z)$ provides a universal covering $\mathbb{C} \to (\mathbb{C}, \nu)$. Show that the Euler characteristic $\chi(\mathbb{C}, \nu)$ is zero, and the fundamental group $\pi_1(S, \nu)$ consists of all transformations of the form $\gamma : z \mapsto n \pm z$. with $n \in \mathbf{Z}$.

Problem E-2. For $S = \hat{\mathbb{C}}$ with three ramified points $\nu(0) = \nu(1) =$

$\nu(\infty) = 2$, show that the rational map $\pi(z) = -4z^2/(z^2 - 1)^2$ provides a universal covering $\widehat{\mathbb{C}} \to (\widehat{\mathbb{C}}, \nu)$. Show that $\chi(\widehat{\mathbb{C}}, \nu) = 1/2$, that the degree is equal to $\chi(\widehat{\mathbb{C}})/\chi(\widehat{\mathbb{C}}, \nu) = 4$, and that the fundamental group consists of all transformations $\gamma : z \mapsto \pm z^{\pm 1}$.

Problem E-3. For a sphere $\widehat{\mathbb{C}}$ with one ramified point, or with two ramified points with different ramification indices, use E.2 to show that there cannot be a regular branched covering surface $S \to (\widehat{\mathbb{C}}, \nu)$.

Appendix F. No Wandering Fatou Components.

This appendix will outline a proof of the following. (Compare §16.)

Theorem F.1 (Sullivan). *If f is a rational map, then every Fatou component of f is eventually periodic.*

The intuitive idea of the proof is the following. Let U be any Fatou component for f, that is any connected component of $\widehat{\mathbb{C}} \smallsetminus J(f)$. Suppose that we try to change the conformal structure on U. If f is to preserve this new conformal structure, then we must also change the conformal structure everywhere throughout the grand orbit of U in a compatible manner. If $f(U) = U$, then the condition that f preserves this structure imposes very strong restrictions. Similarly, if U is periodic or even eventually periodic, then there are very strong restrictions. However, if U were a *wandering component*, that is if the successive forward images

$$U, \quad f(U), \quad f^{\circ 2}(U), \quad f^{\circ 3}(U), \quad \cdots$$

were pairwise disjoint, then we could change the conformal structure within U in an arbitrary manner, and then propagate this change throughout the entire grand orbit of U. As Sullivan realized, this would be too much of a good thing. He showed that it would yield an infinite dimensional space of essentially different rational maps of the same degree. But this is patently impossible, since a rational map of given degree is completely determined by a finite number of complex parameters.

There are two key difficulties in carrying out this argument. The first is that conformal structures constructed in this way are usually discontinuous at every limit point of the grand orbit of U, so that it is not easy to make sense of them. However, this problem had been dealt with earlier in the pioneering work of Morrey, Ahlfors, and Bers. The second difficulty that Sullivan faced was the need for some effective way of showing that he really got too many distinct rational maps in this way, and not just many new ways of constructing the same old rational maps. I will describe a way of dealing with this second problem by means of cross-ratios.

The Beltrami Equation. Before beginning this argument, we must explain the concept of a measurable conformal structure on an open set $U \subset \mathbb{C}$. Intuitively, a conformal structure at a point $z \in U$ can be prescribed by choosing some ellipse centered at the origin in the tangent space $T_z U \cong \mathbb{C}$. We are to think of this ellipse as a "circle" in the new conformal structure. In more technical language, a *conformal structure* at the point z is determined by a *complex dilatation* $\mu(z) \in \mathbb{D}$. First consider the case where $\mu(z)$ is constant. Then the function $h(z) = z + \overline{z}\mu$ satisfies the *Beltrami differential equation*

$$\frac{\partial h}{\partial \bar{z}} = \mu(z)\frac{\partial h}{\partial z} \qquad \text{(F : 1)}$$

(named for Eugenio Beltrami, 1835-1900). Here the derivatives $\partial/\partial\bar{z}$ and $\partial/\partial z$ are to be defined by the formula*

$$\frac{\partial}{\partial\bar{z}} = \frac{1}{2}\left(\frac{\partial}{\partial x} + i\frac{\partial}{\partial y}\right), \qquad \frac{\partial}{\partial z} = \frac{1}{2}\left(\frac{\partial}{\partial x} - i\frac{\partial}{\partial y}\right),$$

where $z = x + iy$. Note that a round circle $|h| = $ constant in the h-plane corresponds to an ellipse $|z + \bar{z}\mu| = $ constant in the z-plane, with direction of the major axis controlled by the argument of μ, and with eccentricity controlled by $|\mu|$. (If $|\mu| = r < 1$, then the ratio of major axis to minor axis is equal to $(1 + r)/(1 - r)$, which tends to infinity as $r \to 1$.)

More generally, if the function $\mu(z)$ is real analytic, then Gauss, in his construction of "isothermal coordinates", showed that an equation equivalent to (F : 1) always has local solutions. Morrey extended this to the case where $\mu(z)$ is measurable, with

$$|\mu(z)| < \text{constant} < 1, \qquad \text{(F : 2)}$$

constructing a solution $z \mapsto h(z)$ which maps a region in the z plane homeomorphically onto a region in the h plane. Furthermore, if h_1 and h_2 are two distinct solutions, he showed that the composition $h_2 \circ h_1^{-1}$ is holomorphic.

Here some explanation is needed, since we are considering a differential equation involving non-differentiable functions. For any open set $U \subset \mathbb{C}$ let $L^1(U)$ be the vector space consisting of all measurable functions $\phi : U \to \mathbb{C}$ with

$$\iint_U |\phi(x + iy)|\, dx\, dy < \infty.$$

We will also need the vector space of *test functions*, consisting of all C^∞ functions $\tau : U \to \mathbb{C}$ which vanish outside of some compact subset of U.

Definition. A continuous function $h : U \to \mathbb{C}$ has *distributional derivatives in* L^1 if there are elements h_z and $h_{\bar{z}}$ in $L^1(U)$ so that

$$\iint_U \left(h_z(z)\,\tau(z) + h(z)\,\partial\tau/\partial z\right) dx\, dy = 0 \qquad \text{(F : 3)}$$

for every such test function τ; with an analogous equation for $h_{\bar{z}}$. Note that these partial derivatives are only defined almost everywhere: we can

---

*   To illustrate this notation, if $f(z)$ satisfies the Cauchy-Riemann equation $\partial f/\partial\bar{z} = 0$, then f is holomorphic and $\partial f/\partial z$ is the usual holomorphic derivative.

change them on a set of Lebesgue measure zero without affecting (F : 3). The Beltrami equation for h now requires that

$$h_{\bar{z}}(z) \;=\; \mu(z)\,h_z(z)$$

for almost every $z \in U$. This makes sense, since the pointwise product of an L^1 function and a bounded measurable function is again in L^1. By definition, any continuous one-to-one solution h is called a *quasiconformal mapping* on U, with complex dilatation μ.

More generally, we can consider such a measurable conformal structure on a Riemann surface S. However it is no longer described by a complex valued function, but rather by a section of a certain complex line bundle. Given a local coordinate z on an open set U, we can still describe the conformal structure on U by a dilatation function $\mu : U \to \mathbb{D}$, but on the overlap between two local coordinates z and z' the equation

$$\mu'(z') \;=\; \mu(z)\,\frac{\partial z'}{\partial z}\,\frac{\partial \bar{z}}{\partial \bar{z}'}$$

must be satisfied in order to make sense of this structure globally. Note that $|\mu'(z')| \;=\; |\mu(z)|$, so that condition (F : 2) is independent of the choice of coordinate system. If this conformal structure is measurable and satisfies (F : 2) everywhere, then the local solutions h form the atlas of local conformal coordinates for a new Riemann surface S_μ which is topologically identical to S, but conformally (and even differentiably) quite different. In the special case where S is the Riemann sphere, it follows from the Uniformization Theorem that S_μ is conformally equivalent to the Riemann sphere. In particular, there is a unique conformal isomorphism $h : S \to S_\mu$ which fixes the points 0, 1 and ∞. If we remember that S_μ is identical to $S = \hat{\mathbb{C}}$ as a topological space, then we can also describe $h = h_\mu$ as a quasiconformal homeomorphism from $\hat{\mathbb{C}}$ to itself (or briefly a *qc-homeomorphism*) with complex dilatation $\mu(z)$.

We can also study the dependence of h_μ on the dilatation μ. For each fixed z_0, Ahlfors and Bers (1960) showed that the correspondence $\mu \mapsto h_\mu(z_0)$ defines a differentiable function from the appropriate space of dilation functions to the Riemann sphere. For further information, see for example Ahlfors (1987), Carleson and Gamelin, Lehto (1987), or Lehto and Virtanen.

Some Conformal Structures on the Unit Disk. In order to carry out Sullivan's proof, we must construct a large family of essentially distinct conformal structures on the open disk $\mathbb{D}$. This can be done as follows. Let G be the group of all C^∞ diffeomorphisms of the circle $\partial\mathbb{D}$ which fix three specified points, say ± 1 and i. If $e : \mathbb{R}/\mathbb{Z} \to \partial \mathbb{D}$ is the standard diffeomorphism $e(t) = \exp(2\pi i t)$, then writing each group element as

$$g : e(t) \mapsto e(t + v(t))$$

we can identitfy G with a convex set consisting of all C^∞ functions $v : \mathbb{R}/\mathbb{Z} \to \mathbb{R}$ which vanish at three specified points, and such that the correspondence $t \mapsto t + v(t)$ has derivative $1 + v'(t) > 0$ everywhere. *This group G has been constructed so that no g other than the identity map can be extended to a conformal automorphism of the closed disk $\overline{\mathbb{D}}$.* (To see this, note that for each non-identity g we can find four distinct points of $\partial \mathbb{D}$ which map to four points with different cross-ratio.) On the other hand, each $g \in G$ extends to a diffeomorphism $\hat{g}$ of $\overline{\mathbb{D}}$, as follows. If $\eta : [0, 1] \to [0, 1]$ is some fixed smooth function with

$$\eta[0, 1/3] = 0, \qquad \eta[2/3, 1] = 1 ,$$

then g extends to

$$\hat{g}\big(r\,e(t)\big) = r\,e\big(t + \eta(r)v(t)\big) . \qquad\qquad (\text{F} : 4)$$

Evidently this extension depends smoothly on g.

Wandering Components. Suppose that some rational function f has a wandering Fatou component U. Replacing U by some iterated forward image if necessary, we may assume that there are no critical points in the forward images $f^{\circ n}(U)$.

Lemma F.2 (Baker). *If no forward image $f^{\circ n}(U)$, $n \geq 0$, contains a critical point, then U must be simply connected.*

Proof. It is convenient to choose the coordinate so that the point at infinity is in U, all other Fatou components lying in a bounded region of $\mathbb{C}$. Let L be an arbitrary simple closed curve contained in U. Since U is contained in the Fatou set, the collection of iterates $f^{\circ n}$ restricted to U forms a normal family. The area of $f^{\circ n}(U)$ must clearly tend to zero as $n \to \infty$, and it follows that any convergent sequence of iterates must converge locally uniformly to a constant map. Therefore the diameter of the set $f^{\circ n}(L)$ must converge to zero as $n \to \infty$. If B_n is the union of the bounded components of the complement $\mathbb{C} \smallsetminus f^{\circ n}(L)$, then it follows that the diameter of $\overline{B}_n$ also tends to zero as $n \to \infty$. In particular, for n large, no component of $f^{-1}(U)$ can be contained in B_n, so every iterated forward image of $\overline{B}_n$ must be contained in the bounded region $\hat{\mathbb{C}} \smallsetminus U$. This implies that $\overline{B}_n$ is contained in the Fatou set for n large, hence $\overline{B}_n \subset f^{\circ n}(U)$. Therefore $f^{\circ n}(L)$ can be deformed to a point within $f^{\circ n}(U)$. Since $f^{\circ n}$ maps U onto $f^{\circ n}(U)$ by a covering map, it follows that L

can be deformed to a point within U. Thus U is simply connected.   □

Proof of F.1. The proof of Sullivan's Theorem begins as follows. Choose some conformal isomorphism ϕ from U to the unit disk $\mathbb{D}$. With $\hat{g}$ as in (F : 4), we can pull back the conformal structure of $\mathbb{D}$ under the composition $\hat{g} \circ \phi$, and then use f to transport this conformal structure over the entire grand orbit of U. (There may be isolated points in this grand orbit which are pre-critical. The induced conformal structure is not defined at such points, but this will not matter.) For points which are not in the grand orbit of U, we simply use the usual conformal structure. Thus we have described a measurable conformal structure almost everywhere on $\widehat{\mathbb{C}}$, and the condition $|\mu| \leq$ constant < 1 is easily verified. Integrating the resulting Beltrami equation, this yields a family of qc-homeomorphisms h_g (normalized so as to fix three points of the Riemann sphere) and a family of maps f_g so that the following diagram is commutative.

$$
\begin{array}{ccccccc}
\mathbb{D} & \xleftarrow{\phi} & U & \subset & \widehat{\mathbb{C}} & \xrightarrow{f} & \widehat{\mathbb{C}} \\
\downarrow \hat{g} & & \downarrow h_g & & \downarrow h_g & & \downarrow h_g \\
\mathbb{D} & \xleftarrow{\phi_g} & U_g & \subset & \widehat{\mathbb{C}} & \xrightarrow{f_g} & \widehat{\mathbb{C}}
\end{array}
$$

Here U_g is defined to be the image of U under h_g, and the maps on the bottom row are defined in such a way that the diagram is commutative. Since the conformal structures on $\widehat{\mathbb{C}}$ were constructed so as to be invariant under f, it follows that each f_g is holomorphic, and hence is a rational map of the same degree d. (Note that horizontal arrows represent holomorphic maps and vertical arrows represent quasiconformal maps.)

We must show that the rational map f_g depends smoothly on g. Note that a rational map $p(z)/q(z)$ of degree d is uniquely determined by its values on $2d + 1$ distinct points. For a second degree d map $P(z)/Q(z)$ takes the same value as $p(z)/q(z)$ at z_i if and only if the polynomial $p(z)Q(z) - P(z)q(z)$ vanishes at z_i. Furthermore, if such a polynomial equation of degree $2d$ has $2d + 1$ distinct solutions then it must be identically zero. In fact the coefficients of $p(z)$ and $q(z)$, suitably normalized, can be obtained by solving linear equation, and hence depend smoothly on the given data. Now consider the points $h_g(j)$ for $1 \leq j \leq 2d + 1$. Since f_g maps each such point to $h_g(f(j))$, and since both $h_g(j)$ and $h_g(f(j))$ depend smoothly on g, it follows that f_g depends smoothly on g.

Choose some $(2d + 2)$-dimensional submanifold $M_0 \subset G$, and choose some point $g_0 \in M_0$ where the rank of the first derivative of the correspondence $g \mapsto f_g$ from M_0 to the space Rat_d of rational maps takes its maximal value $r \leq 2d + 1$. Then a neighborhood N of g_0 maps smoothly onto an r dimensional submanifold $M_1 \subset \mathrm{Rat}_d$. Taking the pre-image in

N of a regular value in M_1, we obtain a submanifold $M_2 \subset N$ of dimension $2d + 2 - r \geq 1$, with the property that the corresponding maps f_g are all the same. Choose a non-constant path $t \mapsto g(t)$ within M_2, so as to obtain a one-parameter family of qc-homeomorphisms $h'_t = h_{g(t)} \circ h_{g(0)}^{-1}$ which conjugate $f_{g(0)}$ to $f_{g(t)} = f_{g(0)}$. *In other words, each h'_t must commute with $f_{g(0)}$.* It follows that each h'_t must restrict to the identity map on the Julia set $J(f_{g(0)})$, since the periodic points of $f_{g(0)}$ cannot move under a deformation which commutes with it.

This leads to a contradiction as follows. Suppose first, to simplify the discussion, that U is bounded by a Jordan curve. For every four points of ∂U we can define the *cross-ratio relative to $\overline{U}$* by choosing a conformal isomorphism $U \to \mathbb{D}$, extending continuously to the boundary (see 17.16), and then taking the usual cross-ratio in $\mathbb{D}$. But if $g(t) \neq g(0)$ then we can choose four points of the unit circle whose cross-ratio definitely changes under the composition $g(t) \circ g(0)^{-1}$. Hence h'_t cannot fix the corresponding four points of $\partial U_{g(0)}$, contradicting our previous statement. This proves F.1 under the hypothesis that ∂U is a Jordan curve.

If ∂U is not a Jordan curve, then we must elaborate this argument using ideas from §17. Recall that the *Carathéodory compactification* $\hat{U}$ consists of U together with a circle of ideal points which are called *prime ends*. These are constructed using only the topology of the pair $(\overline{U}, \partial U)$. Recall also that any Riemann map $U \to \mathbb{D}$ extends to a homeomorphism $\hat{U} \to \overline{\mathbb{D}}$. Thus, if we are given four distinct prime ends in $\partial \hat{U}$, it follows that the cross-ratio of the corresponding points in $\partial \mathbb{D}$ is well defined, independent of the particular choice of Riemann map. To complete the proof, we will need the following supplementary statement.

Lemma F.3. *If an orientation preserving homeomorphism h from the pair $(\overline{U}, \partial U)$ to itself restricts to the identity map on ∂U, then h maps each prime end of $(\overline{U}, \partial U)$ to itself.*

(The example of the complex conjugation map on the pair $(\hat{\mathbb{C}}, [0, 1])$ shows that some restriction, such this orientation condition, is necessary.) To prove F.3, recall from §17 that a prime end is determined by a fundamental chain $\{A_j\}$ of transverse arcs, with associated neighborhoods $N(A_1) \supset N(A_2) \supset \cdots$. If the corresponding neighborhoods $h(N(A_j))$ were disjoint from the $N(A_j)$, then each $\overline{N(A_j)} \cup h(\overline{N(A_j)})$ would be a region bounded by a Jordan curve. The homeomorphism h must preserve orientation on this region, yet reverse orientation on its boundary, which is impossible. $\square$

Making use of this result, the proof of F.1 goes through as before. $\square$

Appendix G. Parameter Space

A very important part of complex dynamics, which has barely been mentioned in these notes, is the study of parametrized families of mappings. (Compare Figures 25, 26.) As an example, consider the family of all quadratic polynomial maps. A priori, a quadratic polynomial is specified by three complex parameters; however any such polynomial can be put into the unique normal form

$$f(z) \; = \; z^2 + c \qquad\qquad (G:1)$$

by an affine change of coordinates. (Other normal forms which have been used are $\omega \;\mapsto\; \omega^2 + \lambda\omega$, with a preferred fixed point of multiplier λ at the origin, or

$$w \;\mapsto\; \lambda w(1-w) \qquad\qquad (G:2)$$

which is more or less equivalent provided that $\lambda \neq 0$. Here $4c = \lambda(2-\lambda)$ and $\omega = z - \lambda/2 = -\lambda w$.) Using such a normal form, we can make a computer picture in the *parameter space* consisting of all complex constants c or λ . Each pixel in such a picture, corresponding to a small square in the parameter space, is to be assigned some color, perhaps only black or white, which depends on the dynamics of the corresponding quadratic map.

The first crude pictures of this type were made by Brooks and Matelski, as part of a study of Kleinian groups. They used the normal form (G : 1), and introduced the open set consisting of all points of the c-plane for which the corresponding quadratic map has an attracting periodic orbit in the finite plane. I will use the notation $\mathcal{H}$ (standing for "hyperbolic") for this Brooks-Matelski set. At about the same time, Hubbard (unpublished) made much better pictures of a quite different parameter space arising from Newton's method for cubics. Mandelbrot, perhaps inspired by Hubbard, made corresponding pictures for quadratic polynomials, using the normal form (G : 2) and also a variant of (G : 1). In order to avoid confusion, let me translate all of Mandelbrot's definitions to the normal form (G : 1). He introduced two different sets, which I will call Q and M . (Mandelbrot did not give these sets different names, since he believed that they were identical.) By definition, a parameter value c belongs to Q if the corresponding filled Julia set contains an interior point, and belongs to M if its filled Julia set contains the critical point $z = 0$ (or equivalently is connected). The Brooks-Matelski set satisfies $\mathcal{H} \subset Q \subset M$. Mandelbrot made quite good computer pictures, which seemed to show a number of isolated "islands". Therefore, he conjectured that Q [or M] has many distinct connected components. (The editors of the journal thought that his islands were specks of dirt, and carefully removed them from the pictures.) Mandelbrot also described an important smaller set $\mathcal{L} \subset Q$ which he believed to be the principal connected component of Q. This set $\mathcal{L}$ consists of a central cardioid $\mathcal{L}_0$ with some (but not all) boundary points included, together with countably many smaller nearly-round disks which are pasted on inductively, in an explicitly described pattern.

Although Mandelbrot's statements in this first paper were not completely right, he deserves a great deal of credit for being the first to point out the extremely complicated geometry associated with the parameter space for quadratic maps. His major achievement has been to demonstrate to a very wide audience that such complicated "fractal" objects play an important role in a number of mathematical sciences.

The first real mathematical breakthrough came with Douady and Hubbard's work in 1982. They introduced the name *Mandelbrot set* for the compact set M described above, and provided a firm foundation for its mathematical study, proving for example that M is connected with connected complement. (Meanwhile, Mandelbrot had decided empirically that his isolated islands were actually connected to the mainland by very thin filaments.) Already in this first paper, they showed that each hyperbolic component of the interior of M can be canonically parametrized, and showed that the boundary ∂M can be profitably studied by following external rays.

It may be of interest to compare the three sets $\mathcal{H} \subset Q \subset M$ in parameter space. They are certainly different since $\mathcal{H}$ is open, M is compact, and Q is neither. Using Sullivan's work (§16), we can say that Q consists of $\mathcal{H}$ together with a very sparse set of boundary points, namely those for which the corresponding map has either a parabolic orbit or a Siegel disk. Quite likely, there is no difference between these three sets as far as computer graphics are concerned, since it is widely conjectured[*] that the Brooks-Matelski set $\mathcal{H}$ is equal to the interior of M, and that M is equal to the closure of $\mathcal{H}$. However, as far as practical computing is concerned, it should be noted that it is quite easy to test (at least roughly) whether a parameter value belongs to M, but somewhat harder to decide whether it belongs to $\mathcal{H}$ (compare 8.6), and very difficult to decide whether it belongs to Q. (Compare Appendix H. Here I am only speaking of approximate tests. To decide precisely whether a given point belongs to M or whether a given point belongs to $\mathcal{H}$ may be very difficult. As a

[*]   Douady and Hubbard have shown that these conjectures are true if the set M is locally connected. The work of Yoccoz lends support to the belief that M is indeed locally connected. (Compare Hubbard 1993.)

specific example, as noted in the discussion of 8.6, I have no idea how one could decide whether or not the point $c = -1.5 \in M$ belongs to $\mathcal{H}$.)

Another important development came in 1983, with the work of Mañé, Sad and Sullivan on stability of the Julia set $J(f)$ under deformation of f . (Compare the discussion in §19.) These results were obtained independently by Lyubich 1983, 1990. The study of parameter space for higher degree polynomials began some five years later with the work of Branner and Hubbard. Using the normal form

$$f(z) = z^3 - 3a^2 z + b,$$

with the two critical points at $z = \pm a$, they proved that the *cubic connectedness locus*, consisting of all parameter pairs (a, b) for which $J(f)$ is connected, is a *cellular set*. (Compare Problem 9-c.) In particular, this set is compact and connected. A corresponding result for polynomials of higher degree has been obtained by Lavaurs. Parameter space studies for rational maps are more awkward, since there is no obvious normal form. However, an important beginning has been made by Rees. (See also Milnor 1993.) All of these studies are very new, and much remains to be done.

Here are two problems for the reader.

Problem G-1. Show that every polynomial map of degree $d \geq 2$ is conjugate, under an affine change of coordinates, to one in the "Fatou normal form"

$$f(z) = z^d + a_{d-2} z^{d-2} + \cdots + a_1 z + a_0 .$$

Let $P(d) \cong \mathbb{C}^{d-1}$ be the space of all such maps. Show that the cyclic group $Z(d-1)$ of $(d-1)$-st roots of unity acts on $P(d)$ by linear conjugation, replacing $f(z)$ by $f(\omega z)/\omega$, and show that the quotient $P(d)/Z(d-1)$ can be identified with the "moduli space" of degree d polynomials up to affine conjugation. If $d \geq 4$, show that this moduli space is not a manifold.

Problem G-2. Show that every quadratic rational map is conjugate, under a fractional linear change of coordinates, for example to one in the form

$$f(z) = 1 + (z^2 - 1)/(az^2 - b)$$

where $a \neq b$, with critical points at zero and ∞ and a fixed point at $z = 1$. (In general this form is not unique, since such a map actually has three fixed points, and since the roles of zero and infinity can be interchanged.)

Appendix H. Remarks on Computer Graphics

In order to make a computer picture of some complicated compact set $L \subset \mathbb{C}$, for example a Julia set, we must compute a matrix of small integers, where the (i,j)-th entry describes the color (perhaps only black or white) which is assigned to the (i,j)-th "pixel" of the computer screen. Each pixel represents a small square in the complex plane, and the color which is assigned must tell us something about intersection of L with this square.

In the case of a quadratic Julia set $J(f)$, one very fast method involves following iterates of the inverse map f^{-1}, taking all possible branches. (Compare 4.10.) As Mandelbrot points out, this method yields an excellent picture of the outer parts of the $J(f)$ but shows very little detail in the inner parts. If we think of the electrostatic field produced by an electric charge on $J(f)$, this method will emphasize only those parts of the Julia set at which "lines of force" (or "external rays" in the language of §18) tend to land.

A slower but much better procedure for plotting $J(f)$ involves iterating the map f for some large number (perhaps 50 to 50000) of times, starting at the midpoint of each pixel. If the orbit "escapes" from a large disk after n iterations, then the corresponding pixel is assigned a color which depends on n. In more refined versions of this method, one computes not only the value of the n-th iterate of f but also the absolute value of its derivative. Compare the discussion below. Similar remarks apply to the *Mandelbrot set M*, as defined by Douady and Hubbard. (Compare Appendix G). In this case one takes the quadratic map corresponding to the midpoint of the square and follows the orbit of its critical point.

Remark. In order to understand some of the limitations of this method, consider the situation near a fixed point $f(z_0) = z_0$ in the Julia set. First consider the repelling case, with multiplier say of absolute value two. If we start at at point z at distance $1/1000$ from z_0, then the distance from z_0 will roughly double with each iteration. Hence, after only ten iterations the image of z will move substantially away from z_0. The result will be a computer picture which is quite sharp and accurate near z_0. (Figures 7, 8.)

Now suppose that we try to construct a picture for $z \mapsto z + z^4$ by the same method. Again start with a point $z \in J$ at a distance of $\epsilon \approx 1/1000$ from the fixed point at zero. Examining the proof of 7.2, we see that the associated coordinate $w_0 = -1/3z_0^3$, which increases by one under each iteration, is roughly equal to $-1/3\epsilon^3 \approx 300,000,000$. *Thus we would have to follow such an orbit for some $300,000,000$ iterations in order to escape from a neighborhood of $z = 0$.* The result in practice will be a false picture which shows everything near the origin to be in the Fatou set. (This difficulty was eliminated in Figure 18 by using a special computer program, which extrapolated iterates of f in order to distinguish different Fatou components.

For the fixed points of Cremer type, the situation is much worse. As far as I know, no useful computer picture of such a point has ever been produced!

Many Julia sets are made up of very fine filaments. For such sets, it is essential to make some kind of distance estimate in order to obtain a sharp picture. In particular, if the filled Julia set has measure zero then all of the center points of our pixels will quite likely correspond to escaping orbits. But a good distance estimate can tell us that our pixel intersects the set $J(f)$, even though its center point is outside. In the case of the Mandelbrot set M, this procedure is even more important, since M contains both large regions and also very fine filaments. Indeed, it was precisely the difficulty of seeing such filaments which led to Mandelbrot's initial belief that M has many components.

Here is an example of how first derivatives can be used to make such distance estimates. (Compare Milnor 1989, as well as Fisher.) Consider a rational map $f : \hat{\mathbb{C}} \to \hat{\mathbb{C}}$ with a superattractive fixed point at the origin. Let U be the basin of attraction of this fixed point. Let us assume, to simplify the discussion, that this basin is connected, simply connected, and contains no other critical point of f. Then it is not difficult to show that the Böttcher coordinate of §9 can be defined throughout U, and yields a conformal isomorphism $\phi : U \to \mathbb{D}$ with $\phi(f(z)) = \phi(z)^n$. Define the Green's function $G : U \smallsetminus \{0\} \to \mathbb{R}$ by the formula

$$G(z) = \log |\phi(z)| < 0.$$

(Compare §9 and §18.) Denoting the gradient vector of G by G', we will show that:

(1) *the function G and the norm $\|G'\| = |\phi'(z)/\phi(z)|$ are easy to compute, and*

(2) *the distance of z from the boundary of U can be computed, up to a factor of two, from a knowledge of G and $\|G'\|$.*

In fact, for any orbit $z_0 \mapsto z_1 \mapsto \cdots$ in U, it is easy to check that

$$G(z_0) = \lim_{k \to \infty} \log |z_k|/n^k.$$

Since the convergence is locally uniform, we can also write

$$\|G'(z_0)\| = \lim_{k\to\infty} \frac{|dz_k/dz_0|}{n^k|z_k|} .$$

In both cases, the successive terms can easily be computed inductively, and we obtain good approximations by iterating until $|z_k|$ is small. (If many iterations do not yield any small z_k, then we must assume that $z_0 \neq U$, and set $G = 0$.)

Setting $\phi(z) = w$, a brief computation shows that the Poincaré metric on U can be written as

$$\frac{2|dw|}{1-|w|^2} = \frac{2|\phi'(z)dz|}{1-|\phi(z)|^2} = \frac{\|G'(z)dz\|}{|\sinh G(z)|} .$$

As an immediate consequence of the Quarter Theorem A.7 and A.8, we obtain the following.

Corollary. *The distance between z and the boundary of U is equal to $|\sinh(G)|/\|G'\|$, up to a factor of two.*

If z is very close to ∂U, then G is small and this distance estimate is very close to the ratio $|G|/\|G'\|$. It is interesting to note that this is just the step size which would be prescribed if we tried to solve the equation $G(z) = 0$ by Newton's method.

There are similar estimates in the case of a superattractive fixed point at infinity, or in other words for the basin $\mathbb{C} \smallsetminus K(f)$ of a polynomial map. For further information, the reader is referred to Fisher and to Peitgen.

REFERENCES

L. Ahlfors (1966), "Complex Analysis", McGraw-Hill.

L. Ahlfors (1973), "Conformal Invariants", McGraw-Hill.

L. Ahlfors (1987), "Lectures on Quasiconformal Mappings," Wadsworth.

L. Ahlfors and L. Bers (1960), *Riemann's mapping theorem for variable metrics,* Annals of Math. **72**, 385-404.

L. Ahlfors and L. Sario (1960), "Riemann Surfaces", Princeton U. Press.

D. S. Alexander (1994), "A History of Complex Dynamics from Schröder to Fatou and Julia", Vieweg.

V. Arnold (1965), *Small denominators I, on the mappings of the circumference into itself*, Amer. Math. Soc. Transl. (2) **46**, 213-284.

M. Atiyah and R. Bott (1966), *A Lefschetz fixed point formula for elliptic differential operators*, Bull.A.M.S. **72**, 245-250.

I. N. Baker (1968), *Repulsive fixedpoints of entire functions*, Math. Zeit. **104**, 252-256.

I. N. Baker (1976), *An entire function which has wandering domains*, J. Austral. Math. Soc. **22**, 173-176.

A. Beardon (1984), "A Primer on Riemann Surfaces", Cambridge U. Press.

A. Beardon (1991), "Iteration of Rational Functions", Grad. Texts Math. **132**, Springer.

E. Bedford (1990), *Iteration of polynomial automorphisms of* $\mathbb{C}^2$, Preprint, Purdue.

W. Bergweiler (1993), *Iteration of meromorphic functions*, Bull. Amer. Math. Soc. **29**, 151-188.

L. Bieberbach (1916), *Über die Koeffizienten derjenigen Potenzreihen welche eine schlichte Abbildung des Einheitskreises vermitteln*, S.-B. Preuss. Akad. Wiss., 940-955.

P. Blanchard (1984), *Complex analytic dynamics on the Riemann sphere*, Bull. Amer. Math. Soc. **11**, 85-141.

P. Blanchard (1986), *Disconnected Julia sets,* pp. 181-201 of "Chaotic Dynamics and Fractals", edit. Barnsley and Demko, Academic Press.

P. Blanchard and A. Chiu (1990), "Conformal Dynamics: an Informal Discussion", Lecture Notes, Boston University.

P. Bleher and M. Lyubich (1991), *The Julia Sets and Complex Singularities in Hierarchical Ising Models*, Comm. Math. Phys. **141**, 453-474.

L. E. Böttcher (1904), *The principal laws of convergence of iterates and their application to analysis* (Russian), Izv. Kazan. Fiz.-Mat. Obshch.

14, 155-234.

L. de Brange (1985), *A proof of the Bieberbach conjecture,* Acta Math. **154**, 137-152.

B. Branner (1986), *The parameter space for complex cubic polynomials,* pp. 169-179 of "Chaotic Dynamics and Fractals", edit. Barnsley and Demko, Academic Press.

B. Branner (1989), *The Mandelbrot set,* pp. 75-105 of "Chaos and Fractals", edit. Devaney and Keen, Proc. Symp. Applied Math. **39**, Amer. Math. Soc.

B. Branner and J. H. Hubbard (1988), *The iteration of cubic polynomials, Part I: the global topology of parameter space,* Acta Math. **160**, 143-206.

B. Branner and J. H. Hubbard (1992), *The iteration of cubic polynomials, Part II: patterns and parapatterns,* Acta Math. **169**, 229-325.

H. Brolin (1965), *Invariant sets under iteration of rational functions,* Arkiv för Mat. **6**, 103-144.

R. Brooks and P. Matelski (1978), *The dynamics of 2-generator subgroups of $PSL(2, \mathbb{C})$,* pp. 65-71 of "Riemann Surfaces and Related Topics", Proceedings 1978 Stony Brook Conference, edit. Kra and Maskit, Ann. Math. Stud. **97** Princeton U. Press 1981.

M. Brown (1960), *A proof of the generalized Schoenflies theorem,* Bulletin A.M.S. **66**, 74-76. (See also: *The monotone union of open n-cells is an open n-cell,* Proc.A.M.S. **12** (1961) 812-814.)

H. Bruin, G. Keller, T. Nowicki, S. van Strien (1996), *Wild Cantor attractors exist,* Annals of Math **143**, 97-130.

A. D. Bryuno (1965), *Convergence of transformations of differential equations to normal forms,* Dokl. Akad. Nauk USSR **165**, 987-989.

C. Camacho (1978), *On the local structure of conformal mappings and holomorphic vector fields,* Astérisque **59-60**, 83-94.

C. Carathéodory (1913), *Über die Begrenzung einfach zusammenhängender Gebiete,* Math. Ann. **73**, 323-370. (Gesam. Math. Schr., v. 4.)

L. Carleson and T. Gamelin (1993), "Complex Dynamics", Springer.

A. Cayley (1879), *Application of the Newton-Fourier method to an imaginary root of an equation,* Quart. J. Pure Appl. Math. **16**, 179-185.

E. Coddington and N. Levinson (1955), "Theory of Ordinary Differential Equations", McGraw-Hill.

I. Cornfeld, S. Fomin, Y. Sinai, "Ergodic Theory", Springer 1982.

R. Courant and H. Robbins (1941), "What is Mathematics?", Oxford U. Press.

H. Cremer (1927), *Zum Zentrumproblem,* Math. Ann. **98**, 151-163.

H. Cremer (1938), *Über die Häufigkeit der Nichtzentren,* Math. Ann. **115**, 573-580.

A. Denjoy (1926), *Sur l'itération des fonctions analytiques,* C. R. Acad. Sci. Paris **182**, 255-257.

A. Denjoy (1932), *Sur les courbes définies par les équations différentielles à la surface du tore,* Journ. de Math. **11**, 333-375.

R. Devaney (1986), *Exploding Julia sets,* pp. 141-154 of "Chaotic Dynamics and Fractals", edit. Barnsley and Demko, Academic Press.

R. Devaney (1989), "An Introduction to Chaotic Dynamical Systems", 2nd ed., Addison-Wesley. (Part 3.)

A. Douady (1982-3), *Systèmes dynamiques holomorphes,* Séminar Bourbaki, 35e année, n° 599; Astérisque **105/106** (1983) 39-63.

A. Douady (1986), *Julia sets and the Mandelbrot set,* pp. 161-173 of "The Beauty of Fractals", edit. Peitgen and Richter, Springer.

A. Douady (1987), *Disques de Siegel et anneaux de Herman,* Sém. Bourbaki 39e année (1986-87) n° 677; Astérisque **152-153**, 151-172.

A. Douady and J. H. Hubbard (1982), *Itération des polynômes quadratiques complexes,* C. R. Acad. Sci. Paris **294**, 123-126.

A. Douady and J. H. Hubbard (1984-85), "Étude dynamique des polynômes complexes I & II," Publ. Math. Orsay.

A. Douady and J. H. Hubbard (1985), *On the dynamics of polynomial-like mappings,* Ann. Sci. Ec. Norm. Sup. (Paris) **18**, 287-343.

A. Douady and J. H. Hubbard (1993), *A proof of Thurston's topological characterization of rational functions,* Acta Math. **171**, 263-297.

J. Écalle (1975), *Théorie itérative: introduction a la théorie des invariants holomorphes,* J. Math. Pure Appl. **54**, 183-258.

A. Epstein (1999), *Infinitesimal Thurston Rigidity and the Fatou-Shishikura Inequality,* Stony Brook I.M.S. Preprint 1999#1.*

D. B. A. Epstein (1981), *Prime Ends,* Proc. London Math. Soc. **42**, 385-414.

A. Eremenko and M. Lyubich (1990), *The dynamics of analytic transformations,* Leningr. Math. J. **1**, 563-634.

A. Eremenko and M. Lyubich (1992), *Dynamical properties of some classes of entire functions,* Ann. Inst. Fourier (Grenoble) **42**, 989–1020. (See also Sov. Math. Dokl. **30** (1984) 592-594; Func. Anal. Appl. **19** (1985) 323-324; and J. Lond. Math. Soc. **36** (1987) 458-468.)

---

* Stony Brook preprints are available at www.math.sunysb.edu/preprints.html

K. Falconer (1990), "Fractal Geometry: Mathematical Foundations and Applications", Wiley. (Ch. 14.)

H. Farkas and I. Kra (1980), "Riemann Surfaces", Springer.

P. Fatou (1906), *Sur les solutions uniformes de certaines équations fonctionnelle*, C. R. Acad. Sci. Paris **143**, 546-548.

P. Fatou (1919-20), *Sur les équations fonctionnelles*, Bull. Soc. math. France **47**, 161-271, and **48**, 33-94, 208-314.

P. Fatou (1926), *Sur l'itération des fonctions transcendantes entières*, Acta Math. **47**, 337-370.

Y. Fisher (1989), *Exploring the Mandelbrot set,* pp. 287-296 of The Science of Fractal Images, edit. Peitgen and Saupe, Springer.

Y. Fisher, J. Hubbard and B. Wittner (1988), *A proof of the uniformization theorem for arbitrary plane domains,* Proc. Amer. Math. Soc. **104**, 413-418.

J. Franks (1982), "Homology and Dynamical Systems," Conference Board Math. Sci., Regional Conference **49**, Amer. Math. Soc.

S. Friedland and J. Milnor (1989), *Dynamical properties of plane polynomial automorphisms*, Erg. Th. & Dy. Sy. **9**, 67-99.

L. Goldberg (1992) *Fixed Points of Polynomial Maps, Part I: Rotation Subsets of the Circle,* Ann. Sci. Éc. Norm. Sup. **25**, 679-685 .

L. Goldberg and L. Keen (1986), *A finiteness theorem for a dynamical class of entire functions*, Erg. Th. & Dy. Sy. **6**, 183-192.

L. Goldberg and J. Milnor (1993), *Fixed point portraits of polynomial maps, Part II: Fixed Point Portraits*, Ann. Sci. Éc. Norm. Sup. Paris **26**, 51-98.

T. H. Gronwall (1914-15), *Some remarks on conformal representation*, Ann. of Math. **16**, 72-76.

D. H. Hamilton (1995), Length of Julia curves, Pac. J. Math. **169**, 75-93.

G. H. Hardy and E. M. Wright (1938 etc.), "An Introduction to the Theory of Numbers," Clarendon Press.

M. Herman (1984), *Exemples de fractions rationnelles ayant une orbite dense sur la sphère de Riemann*, Bull. Soc. Math. France **112**, 93-142.

M. Herman (1979), *Sur la conjugation différentiables des difféomorphismes du cercle à les rotations*, Pub. I.H.E.S. **49**, 5-233.

M. Herman (1986), *Recent results and some open questions on Siegel's linearization theorem of germs of complex analytic diffeomorphisms of C^n near a fixed point*, pp. 138-198 of Proc 8th Int. Cong. Math. Phys., World Sci.

J. Hocking and G. Young (1961), "Topology," Addison-Wesley.

K. Hoffman (1962), "Banach Spaces of Analytic Functions," Prentice-Hall.

J. Hubbard (1986), *The Hénon mapping in the complex domain,* pp. 101-111 of Chaotic Dynamics and Fractals, M. Barnsley and S. Demko, (ed.), Academic Press.

J. H. Hubbard (1993), *Local connectivity of Julia sets and bifurcation loci: three theorems of J.-C. Yoccoz,* pp. 467-511 of "Topological Methods in Modern Mathematics,", ed. Goldberg and Phillips, Publish or Perish.

J. L. W. V. Jensen (1899), *Sur un nouvel et important théoreme de la théorie des fonctions,* Acta Math. **22**, 219-251.

G. Julia (1918), *Memoire sur l'iteration des fonctions rationelles,* J. Math. Pure Appl. **8**, 47-245.

L. Keen (1988), *The dynamics of holomorphic self-maps of* $\mathbb{C}^*$, in "Holomorphic Functions and Moduli", edit. Drasin et al., Springer.

L. Keen (1989), *Julia sets,* pp. 57-74 of "Chaos and Fractals, the Mathematics behind the Computer Graphics", edit. Devaney and Keen, Proc. Symp. Appl. Math. **39**, Amer. Math. Soc.

A. Khintchine (1963), "Continued Fractions", Noordhoff.

P. Koebe (1907), *Über die Uniformizierung beliebiger analytischer Kurven,* Nachr. Akad. Wiss. Göttingen, Math.-Phys. Kl. 191-210.

G. Kœnigs (1884), *Recherches sur les integrals de certains equations fonctionelles,* Ann. Sci. Éc. Norm. Sup. (3e ser.) **1** supplém. 1-41.

K. Kuratowski (1958), "Topologie," Warsaw.

S. Lattès (1918), *Sur l'iteration des substitutions rationelles et les fonctions de Poincaré,* C. R. Acad. Sci. Paris **16**, 26-28.

P. Lavaurs (1989), "Systèmes dynamiques holomorphes: explosion de points périodiques paraboliques", Thesis, Univ. Paris-Sud, Orsay.

L. Leau (1897), *Étude sur les equations fonctionelles à une ou plusièrs variables,* Ann. Fac. Sci. Toulouse **11**, E1-E110.

O. Lehto (1987), "Univalent functions and Teichmüller spaces," Springer-Verlag.

O. Lehto (1998) "Mathematics Without Borders, A History of the International Mathematical Union", Springer.

O. Lehto and K. J. Virtanen (1973), "Quasiconformal Mappings in the Plane,", Springer-Verlag.

E. R. Love (1969), *Thomas Macfarland Cherry,* Bull. London Math. Soc. **1** 224-245.

M. Lyubich (1983), *Some typical properties of the dynamics of rational maps,* Russian Math. Surveys **38** (1983) 154-155. (See also Sov. Math. Dokl. **27**, 22-25.

M. Lyubich (1986), *The dynamics of rational transforms: the topological picture,* Russian Math. Surveys **41:4**, 43-117.

M. Lyubich (1987), *The measurable dynamics of the exponential map,* Siber. J. Math. **28**, 111-127. (See also Sov. Math. Dokl. **35** (1987) 223-226.)

M. Lyubich (1990), *An analysis of the stability of the dynamics of rational functions,* Selecta Math. Sovietica **9**, 69-90. (Russian original published in 1984.)

B. Malgrange (1981/82), *Travaux d'Écalle et de Martinet-Ramis sur les systèmes dynamiques,* Sém. Bourbaki, 34e ann., n° 582.

B. Mandelbrot (1980), *Fractal aspects of the iteration of $z \mapsto \lambda z(1-z)$ for complex λ, z,* Annals NY Acad. Sci. **357**, 249-259.

R. Mañé, P. Sad and D. Sullivan (1983), *On the dynamics of rational maps,* Ann. Sci. Éc. Norm. Sup. Paris (4) **16**, 193-217.

B. Maskit (1987), "Kleinian Groups," Grundl. math. Wiss. **287** Springer.

J. Mather (1982), *Topological proofs of some purely topological consequences of Carathéodory's theory of prime ends,* pp. 225-255 of "Selected Studies", edit. T and G. Rassias, North-Holland.

C. McMullen (1987), *Area and Hausdorff dimension of Julia sets of entire functions,* Trans. Amer. Math. Soc. **300**, 329-342.

C. McMullen (1988), *Automorphisms of rational maps,* pp. 31-60 of "Holomorphic Functions and Moduli I", ed. Drasin, Earle, Gehring, Kra & Marden; Springer.

C. McMullen (1994), "Complex Dynamics and Renormalization," Ann. Math. Studies **135**, Princeton U. Press.

C. McMullen (1996). "Renormalization and 3-manifolds which fiber over the circle." Ann. Math Studies **142**, Princeton U. Press.

W. de Melo and S. van Strien (1993), "One Dimensional Dynamics", Springer. (See also de Melo, "Lectures on One-Dimensional Dynamics," 17° Col. Brasil. Mat., IMPA 1990.)

J. Milnor (1975), *On the 3-dimensional Brieskorn manifolds $M(p,q,r)$,* pp. 175-225 of "Knots, Groups, and 3-Manifolds", edit. Neuwirth, Ann. Math. Studies **84**, Princeton U. Press.

J. Milnor (1989), *Self-similarity and hairiness in the Mandelbrot set,* pp. 211-257 of "Computers in Geometry and Topology", edit. Tangora, Lect. Notes Pure Appl. Math. **114**, Dekker (see p. 218 and §5).

J. Milnor (1992a), *Remarks on iterated cubic maps,* Experimental Math. **1**,

5-24.

J. Milnor (1992b), *Hyperbolic components in spaces of polynomial maps*, Stony Brook I.M.S. Preprint 1992#3.

J. Milnor (1992c), *Local Connectivity of Julia Sets: Expository Lectures*, Stony Brook I.M.S. Preprint 1992#11.

J. Milnor (1993), *Geometry and dynamics of quadratic rational maps*, Experimental Math. **2**, 37-83 (see Appendix B).

J. Milnor (1999a), *Periodic Orbits, Externals Rays and the Mandelbrot Set: An Expository Account*, Stony Brook I.M.S. Preprint 1999#3; Astérisque, to appear.

J. Milnor (1999b), *Rational maps with two critical points** (revised version of IMS preprint 1997#10), submitted to Experimental Math.

J. Milnor (1999c), *Pasting together Julia sets, a worked out example of mating**, submitted to Experimental Math.

J. Milnor (1999d), *On cubic polynomials with periodic critical point*, in preparation.

J. Milnor and W. Thurston (1988), *Iterated maps of the interval*, pp. 465-563 of "Dynamical Systems (Maryland 1986-87)", edit. J.C.Alexander, Lect. Notes Math. **1342**, Springer.

M. Misiurewicz (1981), *On iterates of e^z*, Erg. Th. & Dyn. Sys. **1**, 103-106.

P. Montel (1927), "Leçons sur les Familles Normales", Gauthier-Villars.

J. Munkres (1975), "Topology: A First Course", Prentice-Hall.

V. I. Naĭshul' (1983), *Topological invariants of analytic and area preserving mappings* · · ·, Trans. Moscow Math. Soc. **42**, 239-250.

R. Nevanlinna (1967), "Uniformisierung," Springer.

S. Newhouse (1974), *Diffeomorphisms with infinitely many sinks*, Topology **16**, 9-18.

M. Ohtsuka (1970), "Dirichlet Problem, Extremal Length and Prime Ends," van Nostrand.

B. O'Neill (1966), "Elementary Differential Geometry", Academic Press.

H.-O. Peitgen (1988), *Fantastic deterministic fractals*, pp. 169-218 of "The Science of Fractal Images," Barnsley et al., Springer.

R. Perez-Marco (1990), "Sur la dynamique des germes de difféomorphismes holomorphes de $(\mathbf{C}, \mathbf{0})$ et des difféomorphismes analytiques du cercle," Thèse, Paris-Sud.

R. Perez-Marco (1992), *Solution complete au Probleme de Siegel de lineari-*

---

* Available at www.math.sunysb.edu/~jack

sation d'une application holomorphe au voisinage d'un point fixe (d'apres J.-C. Yoccoz), Sem. Bourbaki, n° 753: Astérisque **206**, 273-310.

R. Perez-Marco (1997), *Fixed points and circle maps*, Acta Math. **179**, 243-294.

C. L. Petersen (1993), *On the Pommerenke-Levin-Yoccoz inequality*, Ergodic Theory Dynamical Systems **13**, 785-806.

C. L. Petersen (1996), *Local connectivity of some Julia sets containing a circle with an irrational rotation.* Acta Math. **177**, 163-224.

C. L. Petersen (1998), *Puzzles and Siegel disks*, Progress in Holomorphic Dynamics, 50-85, Pitman Res. Notes Math. Ser., **387**, Longman, Harlow.

G. A. Pfeifer (1917), *On the conformal mapping of curvilinear angles; the functional equation* $\phi[f(x)] = a_1\phi(x)$, Trans. A. M. S. **18**, 185-198.

K. Pilgrim and Tan Lei (1999), *Rational maps with disconnected Julia set*, Astérisque, to appear.

C. Pommerenke (1986), *On conformal mapping and iteration of rational functions*, Complex Var. Th. & Appl. **5**, 117-126.

M. Rees (1984), *Ergodic rational maps with dense critical point forward orbit*, Erg. Th. & Dy. Sy. **4**, 311-322.

M. Rees (1986a), *Positive measure sets of ergodic rational maps*, Ann. Sci. École Norm. Sup. (4) **19**, 383-407.

M. Rees (1986b), *The exponential map is not recurrent*, Math. Zeit. **191**, 593-598.

M. Rees (1990), *Components of degree two hyperbolic rational maps*, Invent. Math. **100**, 357-382.

M. Rees (1992), *A partial description of parameter space of rational maps of degree two, Part 1*, Acta Math. **168**, 11-87.

M. Rees (1995), *A Partial Description of the Parameter Space of Rational Maps of Degree Two: Part 2*, Proc. London Math. Soc. (3) **70**, 644–690.

F. and M. Riesz (1916), *Über die Randwerte einer analytischen Funktion*, Quatr. Congr. Math. Scand. Stockholm, 27-44.

J. F. Ritt (1920), *On the iteration of rational functions*, Trans. Amer. Math. Soc. **21**, 348-356.

P. Roesch (1998), *Topologie locale des méthodes de Newton cubiques: plan dynamique*, C. R. Acad. Sci. Paris S. I Math. **326**, 1221–1226.

D. Schleicher (1997), *Rational parameter rays of the Mandelbrot set*, Stony Brook I.M.S. preprint 1997#13.

E. Schröder (1871), *Ueber iterirte Functionen*, Math. Ann. **3**; see p. 303.

G. Segal (1979), *The topology of spaces of rational functions*, Acta Math.

143, 39-72.

M. Shishikura (1987), *On the quasiconformal surgery of rational functions*, Ann. Sci. Éc. Norm. Sup. **20**, 1-29.

M. Shishikura (1991), *The Hausdorff dimension of the boundary of the Mandelbrot set and Julia sets*, Stony Brook I.M.S. Preprint 1991#7.

M. Shub (1987), "Global Stability of Dynamical Systems," Springer.

C. L. Siegel (1942), *Iteration of analytic functions*, Ann. of Math. **43**, 607-612.

C. L. Siegel and J. Moser (1971), "Lectures on Celestial Mechanics," Springer.

S. Smale (1967), *Differentiable dynamical systems*, Bull. Amer. Math. Soc. **73**, 747-817.

G. Springer (1957), "Introduction to Riemann Surfaces", Addison-Wesley.

N. Steinmetz (1993), "Rational Iteration: Complex Analytic Dynamical Systems", de Gruyter.

D. Sullivan (1983), *Conformal dynamical systems,* pp. 725-752 of "Geometric Dynamics", edit. Palis, Lecture Notes Math. **1007** Springer.

D. Sullivan (1985), *Quasiconformal homeomorphisms and dynamics I, solution of the Fatou-Julia problem on wandering domains*, Ann. Math. **122**, 401-418.

G. Świątek (1998), *On critical circle homeomorphisms*, Bol. Soc. Brasil. Mat. **29**, 329-351.

Tan Lei (1997), *Branched coverings and cubic Newton maps*, Fund. Math. **154**, 207-26.

Tan Lei and Yin Y. (1996), *Local connectivity of the Julia set for geometrically finite rational maps*, Science in China A **39**, 39-47.

W. Thurston (1997), "Three-dimensional Geometry and Topology. Vol. 1", Edited by Silvio Levy. Princeton Mathematical Series **35**, Princeton University Press, Princeton, NJ.

S. Ulam and J. von Neumann (1947), *On combinations of stochastic and deterministic processes*, Bull. Amer. Math. Soc. **53**, 1120.

S. M. Voronin (1981), *Analytic classification of germs of conformal maps $(\mathbb{C}, 0) \to (\mathbb{C}, 0)$ with identity linear part*, Func. Anal. Appl. **15**, 1-17.

G. T. Whyburn (1964), "Topological Analysis," Princeton U. Press.

T. Willmore (1959), "An Introduction to Differential Geometry", Clarendon.

M. Yampolsky (1995), *Complex Bounds for Critical Circle Maps*, Stony Brook I.M.S. Preprint 1995#12.

J.-C. Yoccoz (1984), *Conjugation différentiable des difféomorphismes du cercle dont le nombre de rotation vérifie une condition diophantienne*, Ann. Sci. E.N.S. Paris (4) **17**, 333-359.

J.-C. Yoccoz (1988), *Linéarisation des germes de difféomorphismes holomorphes de* $(\mathbb{C}, 0)$, C. R. Acad. Sci. Paris **306**, 55-58.

E. Zehnder (1977), *A simple proof of a generalization of a theorem by C. L. Siegel*, in "Geometry and Topology III", edit. do Carmo and Palis, Lecture Notes Math. **597**, Springer.

INDEX

From Schröder
to Fatou and Julia

Daniel S. Alexander

**A History of
Complex Dynamics**

From Schröder
to Fatou and Julia

1994. viii, 165 pp. Hardcover. DM 58,00
ISBN 3-528-06520-6

Contents: Schröder, Cayley and
Newton's Method - The Next Wave:
Korkine and Farkas - Gabriel
Koenigs - Iteration in the 1890's:
Grévy - Iteration in the 1890's:
Leau - The Flower Theorem of Fatou
and Julia - Fatou's 1906 Note -
Montel's Theory of Normal Families
- The Contest - Lattès and Ritt -
Fatou and Julia

The contemporary study of complex
dynamics, which has flourished so
much in recent years, is based
largely upon work by G. Julia (1918)
and P. Fatou (1919/20). The goal of
this book is to analyze this work
from an historical perspective and
show in detail, how it grew out of a
corpus regarding the iteration of
complex analytic functions. This
began with investigations by E.
Schröder (1870/71) which he made,
when he studied Newton's method.
In the 1880's, Gabriel Koenigs
fashioned this study into a rigorous
body of work and, thereby,
influenced a lot the subsequent
development. But only, when Fatou
and Julia applied set theory as well
as Paul Montel's theory of normal
families, it was possible to develop
a global approach to the iteration of
rational maps. This book shows,
how this intriguing piece of modern mathematics became reality.

vieweg

Abraham-Lincoln-Straße 46
D-65189 Wiesbaden
Fax 0611. 78 78-400
www.vieweg.de

Stand 1.7.99
Änderungen vorbehalten.
Erhältlich im Buchhandel oder beim Verlag.